Application of Artificial Intelligence in Power System Monitoring and Fault Diagnosis

Application of Artificial Intelligence in Power System Monitoring and Fault Diagnosis

Editors

Guang Wang
Jiale Xie
Shunli Wang

MDPI • Basel • Beijing • Wuhan • Barcelona • Belgrade • Manchester • Tokyo • Cluj • Tianjin

Editors

Guang Wang
North China Electric Power University
Baoding, China

Jiale Xie
North China Electric Power University
Baoding, China

Shunli Wang
Southwest University of Science and Technology
Mianyang, China

Editorial Office
MDPI
St. Alban-Anlage 66
4052 Basel, Switzerland

This is a reprint of articles from the Special Issue published online in the open access journal *Energies* (ISSN 1996-1073) (available at: https://www.mdpi.com/journal/energies/special_issues/AAIPSMFD).

For citation purposes, cite each article independently as indicated on the article page online and as indicated below:

LastName, A.A.; LastName, B.B.; LastName, C.C. Article Title. *Journal Name* **Year**, *Volume Number*, Page Range.

ISBN 978-3-0365-8410-2 (Hbk)
ISBN 978-3-0365-8411-9 (PDF)

Contents

About the Editors

Guang Wang

Guang Wang, Ph.D., is currently an Associate Professor with the Automation Department, North China Electric Power University, Baoding, China. He received a B.E. degree in automation from Harbin Engineering University, Harbin, China, in 2010, and his M.Sc. and Ph.D. degrees in control science and engineering from the Harbin Institute of Technology, Harbin, in 2012 and 2016, respectively. He has led and been involved in more than six research projects, including as the National Natural Science Foundation of China and Natural Science Foundation of Beijing. His research interests include fault diagnosis and performance monitoring, and their applications in energy storage systems.

Jiale Xie

Jiale Xie, Ph.D., is currently a lecturer with the Department of Automation, North China Electric Power University, Baoding, China. He received his B.Eng. degree in automation from Harbin Engineering University, Harbin, China, in 2010, and M.Sc. and Ph.D. degrees in control science and engineering from Harbin Institute of Technology, Harbin, in 2012 and 2018, respectively. He has led and been involved in more than four research projects, including National Natural Science Foundation of China, and Central University Basic Scientific Research Business Expenses Special Funds. His research interests focus on intelligent management and safety control of Li-ion-battery-based energy storage systems.

Shunli Wang

Prof. Dr. Shunli Wang is a leading expert on new energy research, and is the head of DTlab and the leader of the new energy measurement and control research team. He researches modeling and state estimation strategies for high-power lithium-ion batteries. He has undertaken more than 40 projects and registered more than 30 patents, publishing over 100 research papers as well as obtaining 20 awards, including the Young Scholar Award, Science and Technology Progress Award, and University & Enterprise Innovation Talent Team Award, etc. He has developed multiple generation systems, improving the reliability of battery systems and expanding its application fields with significant social and economic benefits.

Editorial

Application of Artificial Intelligence in Power System Monitoring and Fault Diagnosis

Guang Wang [1], Jiale Xie [1,*] and Shunli Wang [2]

1 Department of Automation, North China Electric Power University, Baoding 071003, China; guang.wang@ncepu.edu.cn
2 School of Information Engineering, Southwest University of Science and Technology, Mianyang 621000, China; wangshunli@swust.edu.cn
* Correspondence: tellerxie@ncepu.edu.cn

1. Introduction

Emerging technologies such as artificial intelligence (AI), big data analytics, and deep learning have gained widespread attention in recent years and have demonstrated great potential for application in many industrial fields. In power systems, AI and other technologies are also being used as new and powerful tools to replace traditional techniques for feature modeling, performance control, and fault diagnosis in order to obtain superior results. This Special Issue, "Application of Artificial Intelligence in Power System Monitoring and Fault Diagnosis", aims to introduce the latest advances in this field and discusses the application of AI technology in power system modeling and control, state estimation, performance diagnosis, and prognosis, among other fields.

The scope of this Special Issue includes, but is not limited to, the following:

- Data-based abnormalities analysis of thermal power systems and nuclear power systems;
- Fault diagnosis and prediction of wind turbines based on SCADA data;
- Modeling, monitoring, and diagnosis of waste-to-energy, biomass power, and tidal power systems;
- Data-based fault characteristics analysis of power generation equipment;
- Power equipment health monitoring based on vibration signals, sound signals, image signals, thermal infrared signals, etc.
- Control and performance monitoring of photovoltaic power generation systems;
- Modeling, scheduling, control, and monitoring of microgrid systems;
- SOC estimation, SOH estimation, fault detection, isolation, and localization of lithium battery systems;
- State estimation and performance evaluation of large-scale energy storage systems.

From a total of 24 submissions, 10 research papers were published in this Special Issue, with 14 rejected.

2. Highlights of Published Papers

This section provides a summary of this Special Issue of *Energies*, covering published articles [1–10] which address several topics related to AI technologies in power system performance monitoring.

In [1], Barnabei et al. designed a Supervisory Control and Data Acquisition (SCADA)-based framework for the unsupervised anomaly detection of district heating (DH) network generating units. The framework relies on a multivariate machine learning regression model and then uses a sliding threshold approach for the subsequent processing of the model residuals generated during the testing phase. The system was tested against major failures occurring in gas-fired generating units at the DH plant in Aosta, Italy, and the results showed that the framework can detect anomalies successfully.

Citation: Wang, G.; Xie, J.; Wang, S. Application of Artificial Intelligence in Power System Monitoring and Fault Diagnosis. *Energies* **2023**, *16*, 5477. https://doi.org/10.3390/en16145477

Received: 17 July 2023
Accepted: 18 July 2023
Published: 19 July 2023

In [2], Lin et al. proposed a new method for shunt capacitor monitoring. The method monitors the shunt capacitor bank via the synchronous voltage and branch current of the shunt capacitor bank, calculates the capacitance parameters of the ungrounded star-connected capacitor bank using the parameter symmetry of the capacitor parameter calculation method, and identifies the abnormal state of the capacitor according to the statistical method. The simulation established by PSCAD verified that the relay protection device could effectively monitor the early abnormal condition of the capacitor bank.

In [3], Jawad et al. proposed a fault diagnosis method based on probabilistic generative models to remedy the shortcomings of existing fault detection methods for high-voltage direct current (HVDC) transmission systems. The method uses wavelet transform based on ant colony optimization and artificial neural network to detect different types of faults in HVDC transmission lines. The experimental results showed that the proposed method has higher accuracy and stronger robustness in the fault diagnosis of HVDC transmission systems compared with existing methods, such as support vector machines and decision trees.

In [4], Pujana proposed a hybrid model-based method for developing a digital twin (DT) model for wind power conversion systems. The method combines the advantages of physical models with advanced data analysis techniques to obtain knowledge from actual operational data while preserving physical relationships, thereby generating synthetic data from non-occurring events to detect and classify faults. Compared with existing DT methods, the method proposed in this paper has significant advantages in accuracy and interpretability.

In [5], Xia et al. proposed a multi-model fusion ensemble learning algorithm based on stacked structures to detect power theft. To solve the problem of existing methods being unable to further improve the accuracy of electricity theft detection, a heterogeneous ensemble learning method is used to construct a heterogeneous integrated learning model for stacked structure electricity theft detection using different powerful individual learning superposition integration structures to achieve the accurate detection and identification of electricity theft.

For identifying different types of partial discharges (PDs) in gas-insulated switchgear (GIS), Zheng et al. proposed an improved feature fusion convolutional neural network (IFCNN) method in [6], which solves the problem of traditional methods requiring a large quantity of statistical discharge data. By fusing time-frequency features, the method can uncover more local features of potential discharge pulses and increase the recognition accuracy to 95.8%.

In [7], Luo et al. designed an automatic machine learning-based lifetime prediction model (AutoML) for accurately estimating and predicting the capacity and lifetime of Li-ion batteries. The features of CC and CV phases are extracted using optimized incremental capacity (IC) curves, and the noise is removed using the Kalman filtering algorithm. They then built AutoML, which can automatically generate the appropriate processing flow, addressing the issues of information redundancy and high computational cost. By validating the NASA dataset, they demonstrated a significant improvement in the model's ability to predict battery life on small-scale datasets.

In [8], Bai et al. proposed an HOG-SVM-based power system equipment identification method. First, wavelet transform is performed on the sound signals of power system equipment collected from the field to obtain wavelet coefficient-time maps. Then, the HOG features of the images are selected, and the selected features are classified using an SVM classifier. Moreover, the method also combines sound signal and image processing to effectively take advantage of image processing and avoid the limitations of sound signal processing. Finally, simulation experiments demonstrated that the proposed method can accurately identify and classify power system equipment.

In [9], Chen et al. proposed a deep-learning-based method for the intelligent modeling of the incineration process in waste-to-energy plants. The output variables are selected regarding safety, stability, and economy. The input variables are determined by eliminating invalid redundant variables using the Lasso (Least absolute shrinkage and selection operator) algorithm and a multi-input multi-output model based on feature selection, and CNN-BiLSTM is established. The results showed that the model can fully exploit the data features under multi-dimensional input feature parameters, and that it has higher accuracy and applicability than the traditional model.

Finally, in [10], Zhang et al. constructed a short-term wind speed prediction model based on variable support segments (VSS). At first, the method decomposes the historical wind speed series into several components using the variational mode decomposition method. Then, an improved transformer model is used to predict the predicted values of each element, and these predicted values are summed to obtain the future wind speed prediction. Experimental results showed that the prediction accuracy of the improved transformer model is significantly higher than that of other prediction models.

Author Contributions: Investigation, G.W. and J.X.; Writing—original draft, G.W.; Writing—review and editing, J.X. and S.W. All authors have read and agreed to the published version of the manuscript.

Funding: This work was supported in part by the National Natural Science Foundation of China under Grants 61973117 and 52207235, the Beijing Municipal Natural Science Foundation under Grant 4192056, and the Hebei Natural Science Foundation under Grant F2019502185.

Conflicts of Interest: The authors declare no conflict of interest.

References

1. Barnabei, V.F.; Bonacina, F.; Corsini, A.; Tucci, F.A.; Santilli, R. Condition-Based Maintenance of Gensets in District Heating Using Unsupervised Normal Behavior Models Applied on SCADA Data. *Energies* **2023**, *16*, 3719. [CrossRef]
2. Lin, Y.; Gan, J.; Wang, Z. On-Line Monitoring of Shunt Capacitor Bank Based on Relay Protection Device. *Energies* **2023**, *16*, 1615. [CrossRef]
3. Jawad, R.S.; Abid, H. HVDC Fault Detection and Classification with Artificial Neural Network Based on ACO-DWT Method. *Energies* **2023**, *16*, 1064. [CrossRef]
4. Pujana, A.; Esteras, M.; Perea, E.; Maqueda, E.; Calvez, P. Hybrid-Model-Based Digital Twin of the Drivetrain of a Wind Turbine and Its Application for Failure Synthetic Data Generation. *Energies* **2023**, *16*, 861. [CrossRef]
5. Xia, R.; Gao, Y.; Zhu, Y.; Gu, D.; Wang, J. An Efficient Method Combined Data-Driven for Detecting Electricity Theft with Stacking Structure Based on Grey Relation Analysis. *Energies* **2022**, *15*, 7423. [CrossRef]
6. Zheng, J.; Chen, Z.; Wang, Q.; Qiang, H.; Xu, W. GIS Partial Discharge Pattern Recognition Based on Time-Frequency Features and Improved Convolutional Neural Network. *Energies* **2022**, *15*, 7372. [CrossRef]
7. Luo, C.; Zhang, Z.; Qiao, D.; Lai, X.; Li, Y.; Wang, S. Life Prediction under Charging Process of Lithium-Ion Batteries Based on AutoML. *Energies* **2022**, *15*, 4594. [CrossRef]
8. Bai, K.; Zhou, Y.; Cui, Z.; Bao, W.; Zhang, N.; Zhai, Y. HOG-SVM-Based Image Feature Classification Method for Sound Recognition of Power Equipments. *Energies* **2022**, *15*, 4449. [CrossRef]
9. Chen, L.; Wang, C.; Zhong, R.; Wang, J.; Zhao, Z. Intelligent Modeling of the Incineration Process in Waste Incineration Power Plant Based on Deep Learning. *Energies* **2022**, *15*, 4285. [CrossRef]
10. Zhang, K.; Li, X.; Su, J. Variable Support Segment-Based Short-Term Wind Speed Forecasting. *Energies* **2022**, *15*, 4067. [CrossRef]

MDPI

Article

Variable Support Segment-Based Short-Term Wind Speed Forecasting

Ke Zhang *, Xiao Li and Jie Su

School of Control and Computer Engineering, North China Electric Power University, Baoding 071003, China; 17331250396@163.com (X.L.); sjzjn@126.com (J.S.)
* Correspondence: zhangke_1998@163.com

Abstract: Accurate short-term wind speed forecasting plays an important role in the development of wind energy. However, the inertia of airflow means that wind speed has the properties of time variance and inertia, which pose a challenge in the task of wind speed forecasting. We employ the variable support segment method to describe these two properties. We then propose a variable support segment-based short-term wind speed forecasting model to improve wind speed forecasting accuracy. The core idea is to adaptively determine the variable support segment of the future wind speed by a self-attention mechanism. Historical wind speed series are first decomposed into several components by variational mode decomposition (VMD). Then, the future values of each component are forecast using a modified Transformer model. Finally, the forecasting values of these components are summed to obtain the future wind speed forecasting values. Wind speed data collected from a wind farm were employed to validate the performance of the proposed model. The mean absolute error of the proposed model in spring, summer, autumn, and winter is 0.25, 0.33, 0.31, and 0.29, respectively. Experimental results show that the proposed model achieves significant accuracy and that the modified Transformer model has good performance.

Keywords: wind speed forecasting; variable support segment; VMD; Transformer; attention mechanism

Citation: Zhang, K.; Li, X.; Su, J. Variable Support Segment-Based Short-Term Wind Speed Forecasting. *Energies* **2022**, *15*, 4067. https://doi.org/10.3390/en15114067

Academic Editors: Galih Bangga

Received: 23 April 2022
Accepted: 30 May 2022
Published: 1 June 2022

1. Introduction

Wind energy has become the most promising clean energy due to its large reserves [1] and good foundation. The Global Wind Energy Council has indicated that the installed global wind power capacity provide be up to 20% of global electricity by 2030 [2]. The development and utilization of wind energy are critical to alleviating the pressure generated by traditional energy sources such as fossil fuels. The conversion and management of wind energy is closely related to wind speed. Accurate short-term wind speed forecasting, which estimates the wind speed 30 minutes to 6 hours ahead [3], is essential for optimizing power grid scheduling, reducing system rotating reserve capacity, and guaranteeing stable grid operation [4,5]. However, the accuracy and reliability of wind speed forecasting are affected by the stochastic nature and nonlinear characteristics of wind speed. Various models for improving wind speed accuracy have been proposed [6–9], which can be divided into the categories of single models and combined models based on their structure. The most widely used single models include the backpropagation (BP) neural network [10], extreme learning machine (ELM), Kalman filtering, the autoregressive moving average (ARMA) [11], and support vector regression (SVR) [12] models.

A single model is unable to achieve satisfactory forecasting accuracy due to the intermittency of wind speed. Thus, combined models consisting of multiple single models are widely applied. Extensive studies have shown that combined models have better performance [13,14]. There are two sorts of combined models. The first weights the forecasting values of different models to obtain the final forecasting values. In [15], the weight coefficients of three different models were determined via modified support vector regression. In [16], the partial least squares algorithm was used to optimize the weight coefficients. Wang et al. [17] proposed a combined model in which the coefficients of four artificial neural networks' forecasting results are determined using the multi-objective bat algorithm (MOBA).

However, the original wind speed series often appears as a broadband signal in the frequency domain, which is difficult to forecast directly. Therefore, a second sort of combined model has been presented to solve this issue. First, a historical wind speed series is broken into narrowband components using the signal decomposition method. Then, each narrowband component's future values are forecast separately by the forecasting models. The final forecasting values are the sum of each component's forecasting values. The most commonly used signal decomposition methods include wavelet transform (WT) [18], empirical mode decomposition (EMD) [19] and its variants, and variational mode decomposition (VMD) [20]. In [21], WT was employed to reduce wind speed fluctuation characteristics. Naik et al. [22] utilized EMD to preprocess wind speed data. In [23], VMD was used to overcome the intermittency of the wind and eliminate noise signals. WT requires the wavelet function and the decomposition layers to be selected artificially, which is non-adaptive. Although EMD and its variants are adaptive, they have limitations such as mode mixing and endpoint effect. VMD has good noise robustness, which is an adaptive signal decomposition method. Here, we employ VMD as the signal decomposition method.

Forecasting models are another key component of combined models; research [24,25] has shown that deep learning models have better performance in extracting and learning complex quantitative relationships hidden in wind speed data. Altan et al. [26] used the long short-term memory (LSTM) model for the forecasting of narrowband components, which showed good performance. In [27], the bidirectional LSTM model was utilized to forecast the sub-series. In [28], a combined model which incorporated VMD, differential evolution (DE), and echo state network (ESN) was proposed. In [29], the significant spatiotemporal characteristics in wind speed data were extracted by a graph deep learning model.

The Transformer model [30] is a deep learning model based on the self-attention mechanism which is good at capturing dependencies in long sequences and is not affected by distance. The Transformer model outperforms other deep learning models on process sequence data, hence, we employ it here as the forecasting model. However, the Transformer model cannot be employed for time series forecasting tasks directly due to its particular structure. Therefore, the structure of the Transformer model is modified in this paper. According to the above analysis, we first use VMD to obtain the narrowband components decomposed from historical wind speed series, then utilize the modified Transformer model to obtain each component's forecasting values. The final forecasting values are the sum of each component's forecasting values. The following are this paper's major contributions:

(1) We employ the variable support segment method to describe the time-varying and the inertia properties of wind speed;
(2) We modify the Transformer model in order to approximate the variable support segment and complete the forecasting task of each narrowband component;
(3) We propose a combined model based on the modified Transformer model and VMD. Two evaluation indicators and thirteen baseline models were used for a comparative experiment; the results indicate that our model has higher accuracy than comparative models and that the modified Transformer model outperforms other forecasting models.

The structure of this paper is as follows: Section 2 provides the mathematical description of wind speed forecasting; Section 3 briefly introduces VMD and the Transformer model; Section 4 presents the modified Transformer model and the proposed model; Section 5 analyzes the forecasting results of different models; and the final section contains our conclusions.

2. Mathematical Description of Wind Speed Forecasting

At present, most wind speed forecasting models assume that the future wind speed in the short term is only related to the historical wind speed:

$$\mathbf{x}_N = f(\mathbf{x}_M) \tag{1}$$

where $\mathbf{x}_N = [x_i, \ldots, x_{i+N-1}]$ denotes the future wind speed series and $\mathbf{x}_M = [x_{i-M}, \ldots, x_{i-1}]$ denotes the historical wind speed series (i.e., the support segment of $\mathbf{x}_N$); $f : \mathbb{R}^M \to \mathbb{R}^N$ is the function that describes the mapping relationship between $\mathbf{x}_M$ and $\mathbf{x}_N$. Thus, the wind speed forecasting task can be achieved by constructing a model to approximate the function f. Wind speed series often appear as broadband signals in the frequency domain, while narrowband signals are generally assumed to have a stable future trend and are easier to forecast. As a result, one feasible approach is to forecast the future values based on the narrowband components of historical wind speed series. The wind speed forecasting process based on signal decomposition can be formulated as

$$\mathbf{x}_N = \sum_k \mathbf{x}_N^k = \sum_k f_k(\mathbf{x}_M^k) \tag{2}$$

where $\mathbf{x}_M = \sum_k \mathbf{x}_M^k$, $\mathbf{x}_M^k$ represents the narrowband component of the historical wind speed series, i.e.,the support segment of $\mathbf{x}_N^k$. Therefore, the function $f_k : \mathbb{R}^M \to \mathbb{R}^N$ describes the quantitative relationships between $\mathbf{x}_M^k$ and $\mathbf{x}_N^k$.

The inertia of airflow means that the wind speed shows time-varying and inertial properties, which influences the accuracy of wind speed forecasting. As Equation (2) fails to describe these two properties of wind speed effectively, there is room for improvement. Hence, the parameter τ, which is related to delay, can be introduced to the mathematical description of wind speed forecasting, and the parameter p, which denotes the length of the support segment, is set as a time variable. As a result, the mathematical description of wind speed forecasting can be formulated as

$$\mathbf{x}_N = \sum_k \mathbf{x}_N^k = \sum_k f_k(S_{p_k,\tau_k}^k) \tag{3}$$

where $S_{p_k,\tau_k}^k = [x_{i-p_k-\tau_k}^k, \ldots, x_{i-1-\tau_k}^k]$ is the variable support segment of $\mathbf{x}_N^k$. In Formula (3), the parameters τ and p vary with the historical wind speed series; thus, the inertia property of wind speed is described by the parameter τ, while the time-varying property of wind speed is described by the parameters τ and p jointly. When $N = 1$, Equation (3) corresponds to the one-step wind speed forecasting problem, which can be reformulated as

$$x_i = \sum_k f_k(S_{p_k,\tau_k}^k) \tag{4}$$

Unless otherwise specified, the remainder of this paper concentrates on the issue of one-step wind speed forecasting.

Figure 1 shows the schematic diagram of the variable support segment; $[x_2^1, x_3^1, x_4^1, x_5^1]$, which contributes to the formation of x_{11}^1, is the variable support segment of x_{11}^1, that is, $p_1 = 4$ and $\tau_1 = 5$. Similarly, the variable support segment of x_{11}^2 is $[x_3^2, \ldots, x_7^2]$; $p_2 = 5$ and $\tau_2 = 3$.

According to Equation (4), we can forecast the future wind speed via the following steps.

(1) Decompose the wind speed series into narrowband components based on the signal decomposition method;
(2) Complete the forecasting task of each narrowband component by estimating the variable support segment corresponding to each narrowband component;
(3) Superimpose the forecasting value of each narrowband component to obtain the future wind speed forecasting value.

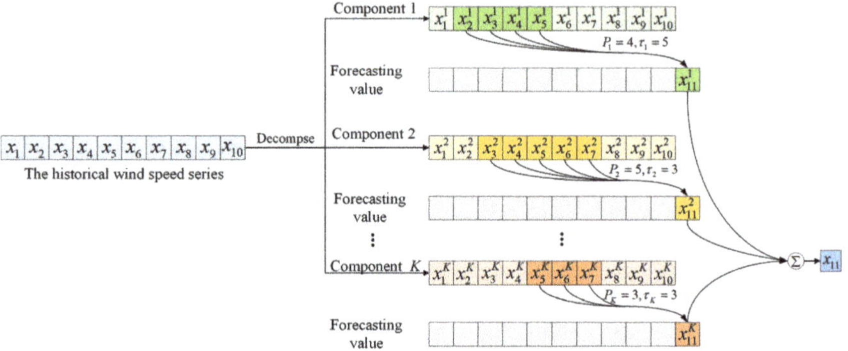

Figure 1. Schematic diagram of the variable support segment.

Approximating the variable support segment accurately is the key to reducing the errors in wind speed forecasting. Existing forecasting models struggle with adaptively approximating the variable support segment. In our approach, the variable support segment is approximated using the self-attention mechanism, the specific process of which is introduced in Section 4.1.

3. VMD and Transformer

For the purposes of this paper, VMD was selected as the signal decomposition method and the Transformer model was selected as the forecasting model; this section briefly introduces them.

3.1. VMD

VMD decomposes an input signal into a number of intrinsic mode functions which are band-limited. It includes two main parts, variational problem construction and variational problem solving.

VMD uses an input signal, $g(t)$, equal to the sum of all the modes as its premise and seeks K mode functions, $u_k(t)$, to obtain the minimum sum of the estimated bandwidths of each mode. Thus, the constrained variational problem can be formulated as

$$\begin{cases} \min\limits_{\{u_k\},\{\omega_k\}} \left\{ \sum\limits_k ||\partial_t[(\delta(t) + \frac{j}{\pi t}) * u_k(t)]e^{-j\omega_k t}||_2^2 \right\} \\ s.t. \sum\limits_k u_k = g(t) \end{cases} \tag{5}$$

where u_k is the mode function, ω_k is the mode center frequency, K is the number of modes, δ is the Dirac disturibution, $*$ is convolution, and $g(t)$ is the input signal.

By introducing the quadratic penalty term α and the Lagrangian multiplier $\lambda(t)$, the constrained variational problem of Equation (5) becomes an unconstrained variational problem:

$$\begin{aligned} &L(\{u_k(t)\}, \{\omega_k\}, \lambda(t)) = \\ &\alpha \sum_k ||\partial_t[(\delta(t) + \frac{j}{\pi t}) * u_k(t)]e^{-j\omega_k t}||_2^2 + ||g(t) - \sum_k u_k(t)||_2^2 + \left\langle \lambda(t), g(t) - \sum_k u_k(t) \right\rangle \end{aligned} \tag{6}$$

In order to solve the unconstrained variational problem, VMD alternately updates $u_k^{n+1}(t)$, ω_k^{n+1}, and $\lambda_k^{n+1}(t)$ to find the "saddle point" of the extended Lagrangian expression. Here, the iterative formula of the Fourier transform of $u_k(t)$, ω_k and $\lambda(t)$ can be expressed as

$$\hat{u}_k^{n+1}(\omega) \leftarrow \frac{\hat{g}(\omega) - \sum_{i \neq k} \hat{u}_i(\omega) + \frac{\hat{\lambda}(\omega)}{2}}{1 + 2\alpha(\omega - \omega_k)^2} \tag{7}$$

$$\omega_k^{n+1}(\omega) \leftarrow \frac{\int_0^\infty \omega |\hat{u}_k(\omega)|^2 d\omega}{\int_0^\infty |\hat{u}_k(\omega)|^2 d\omega} \tag{8}$$

$$\hat{\lambda}^{n+1}(\omega) \leftarrow \hat{\lambda}^n(\omega) + \eta[\hat{g}(\omega) - \sum_k \hat{u}_k^{n+1}(\omega)] \tag{9}$$

where η is an update factor.

3.2. The Transformer Model

The Transformer [30] model is a model based on an "encoder–decoder" structure, shown in Figure 2. The model consists of an input layer, encoder stack, decoder stack, and output layer.

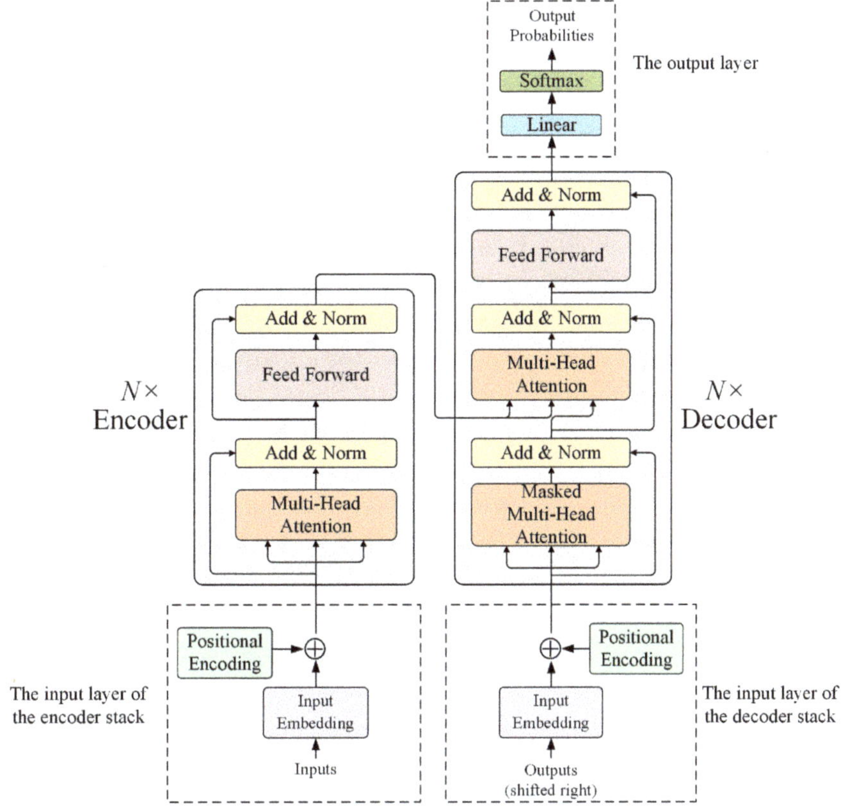

Figure 2. The structure of the Transformer model.

The word embedding module and positional encoding module, which correspond to "Input Embedding" and "Positional Encoding" in Figure 2, respectively, make up the input layer. The word embedding module is utilized to convert input words into computable vectors, as words cannot be directly input into the model. The positional encoding module embeds positional information into the input sequence, as the Transformer model abandons the traditional recurrent neural network structure and is therefore unable to directly receive the position information of the input sequence. The encoder stack which is responsible for encoding the input information and generating intermediate vectors as the input of the decoder stack is composed of several encoders. Each encoder contains two modules, the multi-head attention mechanism module and the feed-forward neural network module, corresponding to "Multi-Head Attention" and "Feed Forward" in the Figure 2, respectively. Here, we use relu as the activation function in the feed-forward neural network module. Residual connections are used between each module and normalization is carried out, which is indicated by the "Add & Norm" part in the Figure 2.

The multi-head attention mechanism module calculates the attention based on a self-attention mechanism, which can deeply explore the internal relationship of input sequences, focus on important information, and filter out unimportant information. The self-attention mechanism first maps the input matrix $\mathbf{X}$ into the query matrix $\mathbf{Q}$, the key matrix $\mathbf{K}$, and the value matrix $\mathbf{V}$, then calculates the attention distribution by the scale dot production, and finally performs a weighted summation of the value matrix according to the attention distribution. Specifically, this is shown in Equations (10)–(13):

$$\mathbf{Q} = \mathbf{W}_Q\mathbf{X} \tag{10}$$

$$\mathbf{K} = \mathbf{W}_K\mathbf{X} \tag{11}$$

$$\mathbf{V} = \mathbf{W}_V\mathbf{X} \tag{12}$$

$$Attention(\mathbf{Q},\mathbf{K},\mathbf{V}) = softmax(\frac{\mathbf{Q}\mathbf{K}^T}{\sqrt{d}})\mathbf{V} \tag{13}$$

where $\mathbf{W}_Q$, $\mathbf{W}_K$, and $\mathbf{W}_V$ are the weight matrix corresponding to $\mathbf{Q}$, $\mathbf{K}$, and $\mathbf{V}$, respectively, and $\sqrt{d}$ is a scale factor.

The information learned by a single self-attention mechanism is relatively simple. In order to fully mine the correlation information between input sequences, the Transformer model further adopts the multi-head attention mechanism in order to learn information from different subspaces, then splices the outputs of different subspaces to obtain the final output, as shown in detail in Equations (14) and (15):

$$Mutilhead(\mathbf{Q},\mathbf{K},\mathbf{V}) = Concat(Head_1,\ldots,Head_H)\mathbf{W}^O \tag{14}$$

$$Head_i = Attention(\mathbf{X}\mathbf{W}_i^Q,\mathbf{X}\mathbf{W}_i^K,\mathbf{X}\mathbf{W}_i^V) \tag{15}$$

where $\mathbf{W}_i^Q$, $\mathbf{W}_i^K$, and $\mathbf{W}_i^V$ are the weight matrices corresponding to $\mathbf{Q}$, $\mathbf{K}$ and $\mathbf{V}$, in $Head_i$, $Concat$ is used to splice the output of each $Head$, and $\mathbf{W}^O$ is the projection matrix, which is used to realize the projection of the stitching result.

The decoder stack, which is responsible for decoding the input information, is composed of several decoders. Compared to the encoder, the decoder includes an additional mask multi-head attention mechanism module to prevent information leakage. Residual connections between the modules of the decoder are used and normalized.

The output layer includes the Linear module and the Softmax module, which are used to convert the vector output by the decoder stack into a probability and then output the word corresponding to the highest probability.

4. The Wind Speed Forecasting Model

4.1. The Modified Transformer Model

The structure of the original Transformer model is not suitable for time series forecasting tasks; therefore, we conducted several specific modifications:

(1) The word-embedding module was replaced by a fully-connected neural network (FCNN) to allow the wind speed series to be input directly into the model;

(2) In the decoder, the masked multi-head attention mechanism was replaced by a multi-head attention mechanism, as only a single data source is fed into the decoder stack and the information of the subsequent sequence is not subsequently involved;
(3) The original output layer was removed and the output of the encoder stack directly mapped into the wind speed forecasting result from the FCNN.

For convenience, the modified Transformer model, shown in Figure 3, is called M-Transformer in this paper.

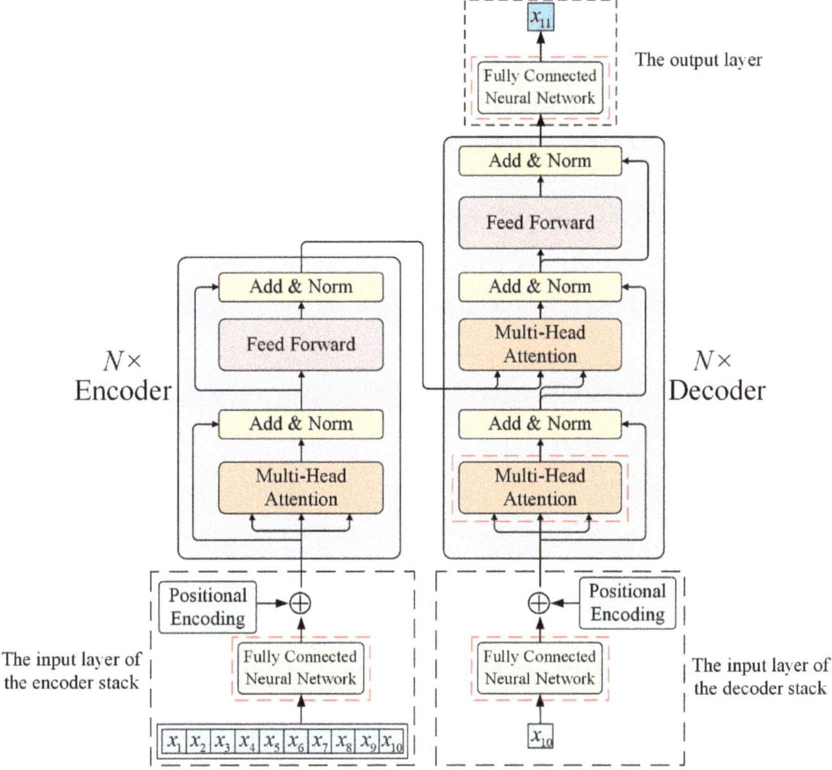

Figure 3. The structure of the M-Transformer model.

Drawing on the large number of previous experimental results, the *Head* number was set as 8 and the input length of the narrowband component as 10. Without loss of generality, the historical wind speed components were represented as $[x_1, \ldots, x_{10}]$. It should be noted that $x_1 \sim x_{10}$ were fed into the FCNN. As shown in Figure 4, x_i is mapped into a row vector by the FCNN with the length $d_s = 512$. The matrix $\mathbf{X}$ is concatenated from ten row vectors generated from the narrowband modes of the historical wind speed, which is then fed into $Head_1 \sim Head_8$ in order to separately calculate the attention distribution.

Using $Head_1$ as an example, the matrix $\mathbf{X}$ is multiplied by $\mathbf{W}_1^Q, \mathbf{W}_1^K, \mathbf{W}_1^V$ to generate $\mathbf{Q}_1, \mathbf{K}_1, \mathbf{V}_1$. The attention distribution of *Head*1 (i.e., the weight matrix $\mathbf{W}_1$ in Figure 4) is calculated based on Equation (16):

$$\mathbf{W}_1 = softmax(\frac{\mathbf{Q}_1\mathbf{K}_1^T}{\sqrt{d}}) \quad (16)$$

after which we multiply $\mathbf{W}_1$ and value matrix $\mathbf{V}_1$ to obtain the output $\mathbf{Z}_1$ of $Head_1$:

$$\mathbf{Z}_1 = \mathbf{W}_1\mathbf{V}_1 \tag{17}$$

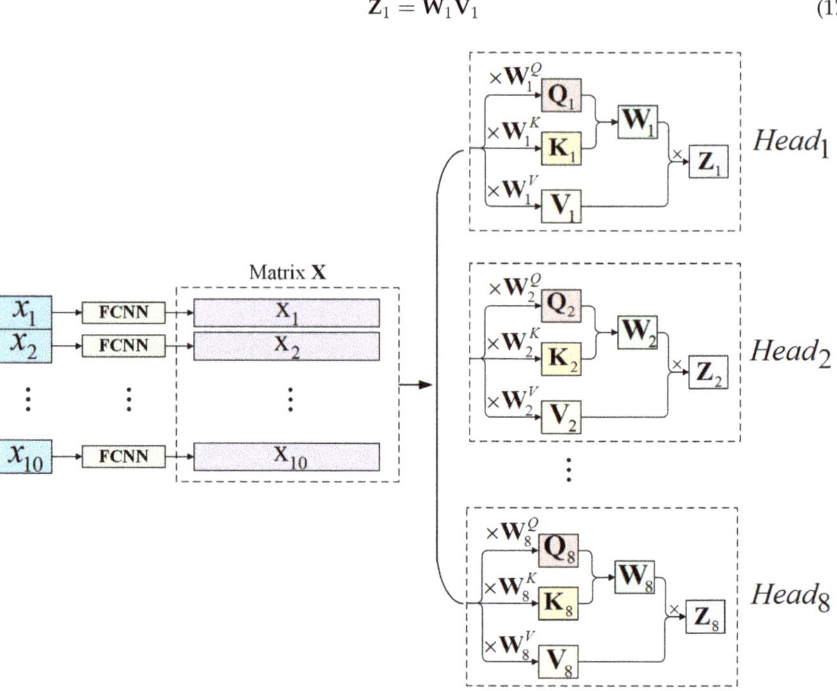

Figure 4. The schematic diagram of the multi-head attention mechanism.

The ith row of $\mathbf{Z}_1$ can be considered as the weighted sum of all rows of the matrix $\mathbf{V}_1$, and the weight of each row is the numerical value of the corresponding element on the ith row of $\mathbf{W}_1$. The jth row of $\mathbf{V}_1$ is determined by the unique historical wind speed component sample value x_j; thus, the weight matrix, $\mathbf{W}_1$, determines which sample values in the narrowband components of the historical wind speed series contribute to the output $\mathbf{Z}_1$ of $Head_1$. Thus, the weight matrices $\{\mathbf{W}_1, \ldots, \mathbf{W}_8\}$ of all $Head$ of the first encoder in the encoder stack together to determine the variable support segment of the narrowband modes of the historical wind speed series, which can be expressed as

$$S_{p,\tau} = \bigcup_{h} \bigcup_{max(\mathbf{W}_h^j)>0} x_j \tag{18}$$

where $\mathbf{W}_h^j$ represents the jth column of the weight matrix $\mathbf{W}_h$ of the hth $Head$ and $max(\mathbf{W}_h^j)$ denotes the maximum element value of $\mathbf{W}_h^j$.

Figure 5 shows a pseudo-color figure of the weight matrix $\mathbf{W}_1$ of the first $Head$ of the first encoder in the encoder stack. It can be seen that the non-zero elements are concentrated in certain columns in $\mathbf{W}_1$, which is to say that in the component of the historical wind speed series, the only elements that contribute to the output of Head are $[x_3, \ldots, x_8]$.

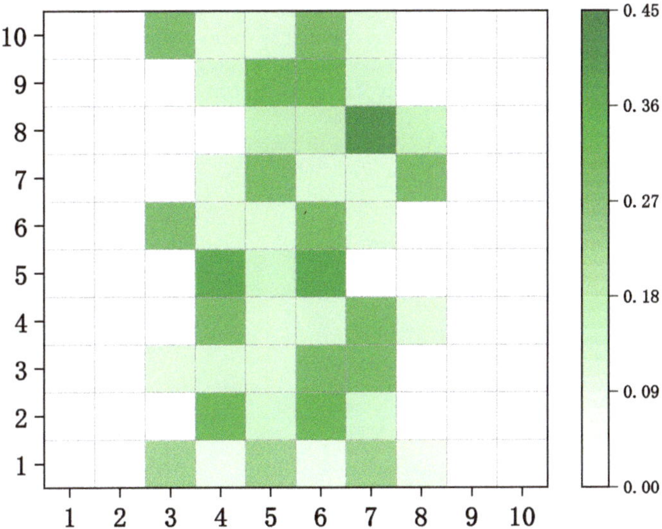

Figure 5. The attention distribution of $Head_1$.

4.2. Proposed Model

According to the wind speed forecasting task steps in Section 2, several narrowband components decomposed from the historical wind speed series are input into the M-Transformer model to separately obtain the forecasting value. The wind speed forecasting result is the sum of the forecasting value of each narrowband component. A flow chart for the proposed method is shown in Figure 6, abbreviated as VMD-TF for convenience.

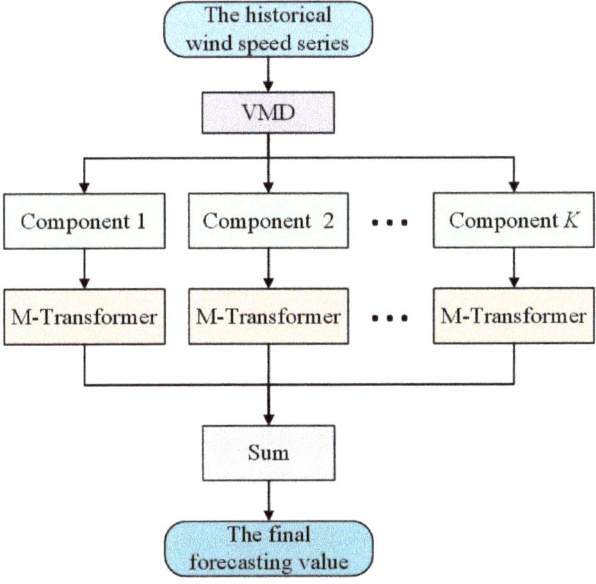

Figure 6. The flowchart of the proposed method.

According to Figure 6, before decomposing the wind speed series based on the VMD, parameters K (i.e., the number of narrowband components) need to be determined. In our approach, these are determined by judging whether the center frequencies of the adjacent components overlap; the specific process is shown in Figure 7. The quadratic penalty factor influences the decomposition results. When the quadratic penalty factor is 2000, VMD has certain adaptability and can avoid mode mixing.

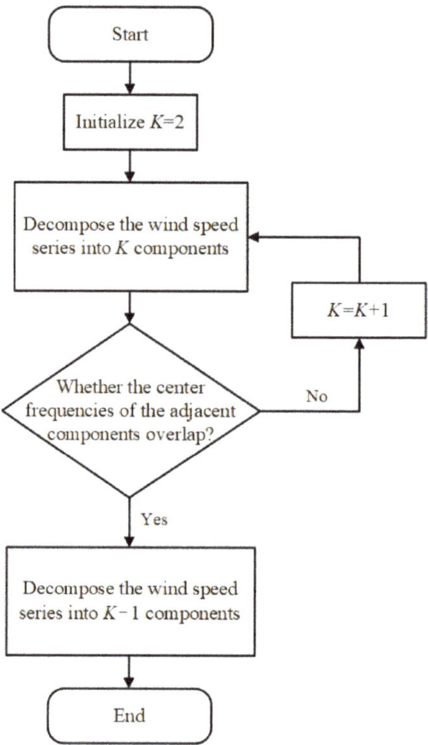

Figure 7. Flowchart for determining K.

5. Experiment and Analysis

5.1. Wind Speed Data

The data were obtained from a wind farm in Hebei. The sampling interval used in collecting the the data was 1 h. Hebei is located in a temperate monsoon climate, and the characteristics of the data consequently vary from season to season. Figure 8 shows the statistics related to the wind speed data in different seasons.

As can be seen in Figure 8, the maximum and average wind speed in summer is higher than in other seasons, indicating abundant wind energy resources. In addition, the wind speed in summer varies greatly and has strong randomness, with the highest standard deviation.

Figure 9 shows the decomposition result of the wind speed series from April 14th to May 18th, that is, in summer, in which C_1–C_7 are narrowband components. It can be clearly seen that the trend of each component is more regular than the original wind speed series. C_8 is the residual component. Although it contains noise, it may contain part of the information of the original wind speed series as well. Therefore, permutation entropy was utilized to assess the signal's randomness and determine whether the residual component could be considered a component of the original wind speed series.

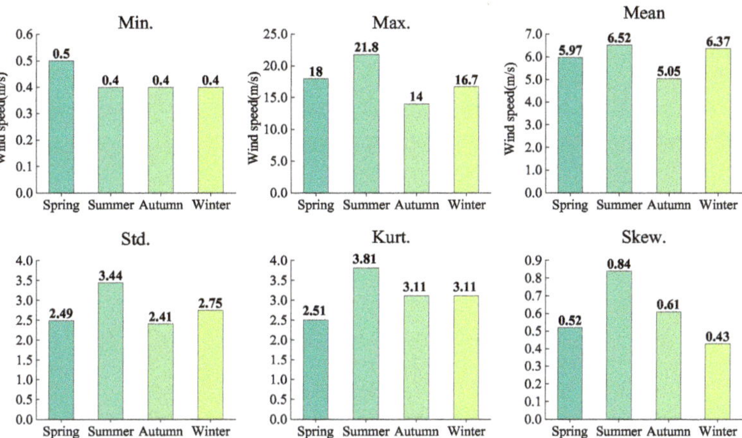

Figure 8. Statistical data for wind speed in different seasons.

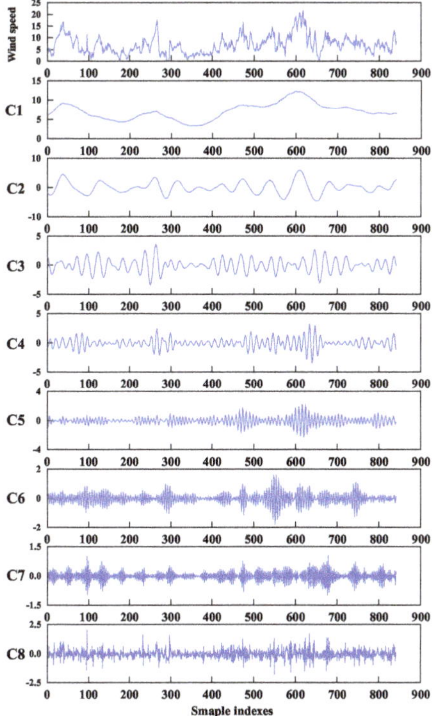

Figure 9. Wind speed series decomposition results.

5.2. Accuracy Assessment

In this paper, the mean absolute error (MAE) and the root mean square error (RMSE) were selected as the evaluation indicators

$$\mathrm{MAE} = \frac{1}{N}\sum_{i=1}^{N}|x_i^a - x_i^f| \tag{19}$$

$$\text{RMSE} = \sqrt{\frac{1}{N}\sum_{i=1}^{N}(x_i^a - x_i^f)^2} \quad (20)$$

where N is the length of the forecasting wind speed series, x_i^a denotes the true value, and x_i^f represents the forecasting value.

5.3. Results and Analysis

5.3.1. Forecasting Result

In each quarter, we randomly selected a week of wind speed data as the test set and used the four weeks of data before the test set as the training set; the specific division is shown in Table 1.

The parameters used for the M-Transformer were as follows: the encoder stack consisted of four encoders, the decoder stack contained four decoders, the dropout rate was equal to 0.1, the learning rate was set to 0.002, the batch size was set to 72 and 1 for the training and testing process, respectively, the optimizer was adma, and the loss function was the mean square error (MSE). Both the training and testing process were implemented in the Python 3.7 platform.

Table 1. Division of the training and testing sets.

Season	Training Set	Testing Set
Spring	25 January–17 February	18 February–23 February
Summer	14 April–11 May	12 May–18 May
Autumn	21 July–17 August	18 August–24 August
Winter	8 September–5 October	6 October–12 October

Figure 10 is a scatter diagram of the forecasting results for each season. The abscissa and ordinate are the forecasting and the true wind speed, respectively. The closer the data points are to the 45° line, the better the forecasting results. In Figure 10, the data points are closely distributed on the 45° line and on both sides, indicating that the proposed model achieves good performance.

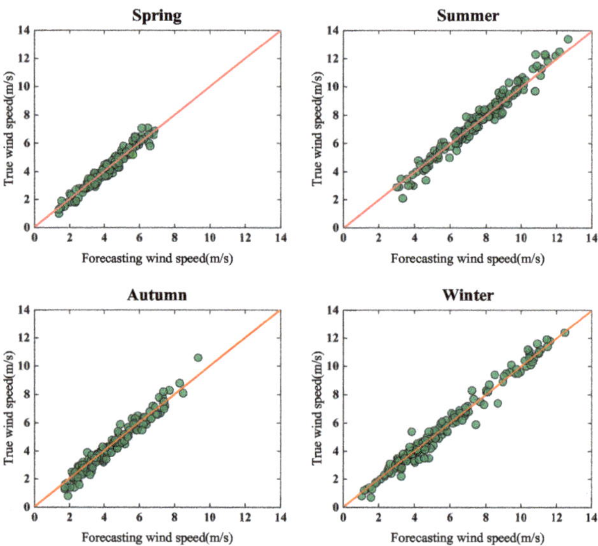

Figure 10. The forecasting results of the proposed model.

5.3.2. Comparative Experiments

In this paper, three single models and three combined models were selected as comparison models to further verify the superiority of VMD-TF. The single models were ARIMA, BP, and LSSVM, and the combined models were EAW [31], WEE [32], and RWA [33]. The forecasting results of each model were evaluated by MAE and RMSE respectively, and the specifics are shown in Figure 11.

Equation (21) was utilized as the evaluation indicator to compare the improvement of VMD-TF to the other models:

$$I_{index} = \frac{E_p - E_c}{E_c} \times 100\% \tag{21}$$

Here, I_{index} denotes the performance improvement index and E_p and E_c are the error of the VMD-TF and the comparison model, respectively. Table 2 shows the specific results.

Table 2. The performance improvements achieved by the proposed model.

		ARIMA	BP	LSSVM	EAW	WEE	RWA
I_{MAE}	Spring	−66%	−61%	−58%	−52%	−48%	−31%
	Summer	−65%	−61%	−58%	−48%	−39%	−30%
	Autumn	−57%	−54%	−52%	−47%	−43%	−28%
	Winter	−65%	−63%	−60%	−45%	−36%	−6%
I_{RMSE}	Spring	−65%	−62%	−59%	−50%	−46%	−31%
	Summer	−57%	−54%	−52%	−42%	−28%	−17%
	Autumn	−63%	−61%	−53%	−49%	−37%	−29%
	Winter	−65%	−61%	−56%	−51%	−42%	−31%

The results of the comparative experiment show the following.

(1) VMD-TF outperforms the other six models. The performance of VMD-TF greatly increased compared with the single models. Using spring as an example, the MAE of VMD-TF fell by 62%, 61%, and 58% compared with ARIMA, BP, and LSSVM, respectively. The reason for this is that the potential of a single model to extract complicated characteristics is limited. However, VMD-TF shows better performance than the three combined models as well. Using autumn as an example, the RMSE of VMD-TF decreased by 49%, 37%, and 21% compared with EAW, WEE, and RWA, respectively, meaning that VMD-TF showed better feature extraction ability than the other combined models.

(2) VMD-TF has the best performance in spring, followed by autumn and winter, and has relatively poor forecasting results in the summer. The properties of the wind speed data in each season have a high relation with the aforementioned results. According to Figure 8, the standard deviation of the summer data are all higher than those in other seasons, indicating that the wind speed in summer fluctuates greatly and is difficult to forecast.

(3) The preceding results illustrate that VMD-TF achieves significant performance. The self-attention mechanism can adjust the attention distribution in a timely fashion according to the input data and realize adaptive estimation of the variable support segment, which is essential for improving wind speed forecasting accuracy.

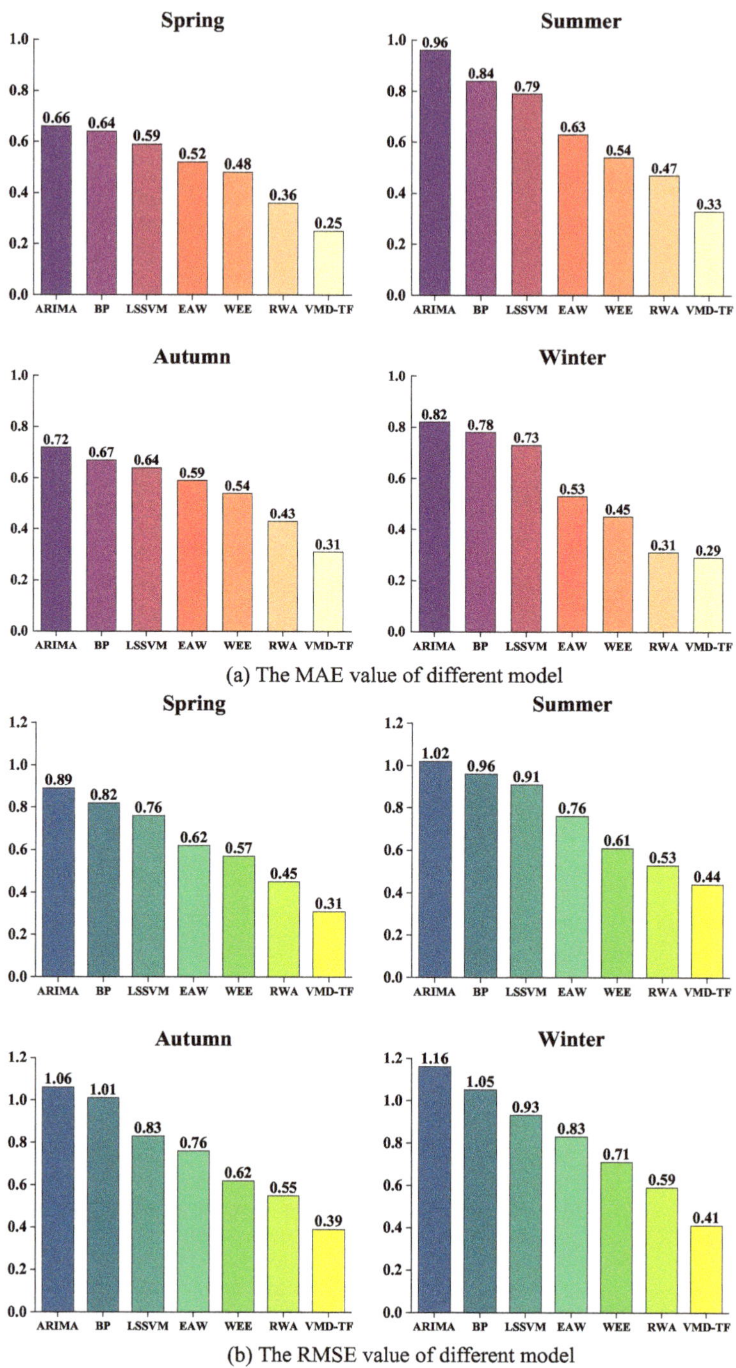

Figure 11. The MAE and RMSE values of different models.

5.3.3. Effectiveness of VMD

We employed EMD and EEMD as comparison methods to demonstrate that VMD could effectively reduce the influence of wind speed non-stationarity. The model combining M-Transformer with EMD is referred to as EMD-TF, while the model combining M-Transformer with EEMD is referred to as EEMD-TF. We used M-Transformer to forecast the wind speed directly without decomposition, in which case it is referred to as TF. In analyzing the capability of these models, the summer testing set was used. Figure 12 exhibits the comparisons between the forecasting values and the true values, while Table 3 shows the forecasting errors for each model.

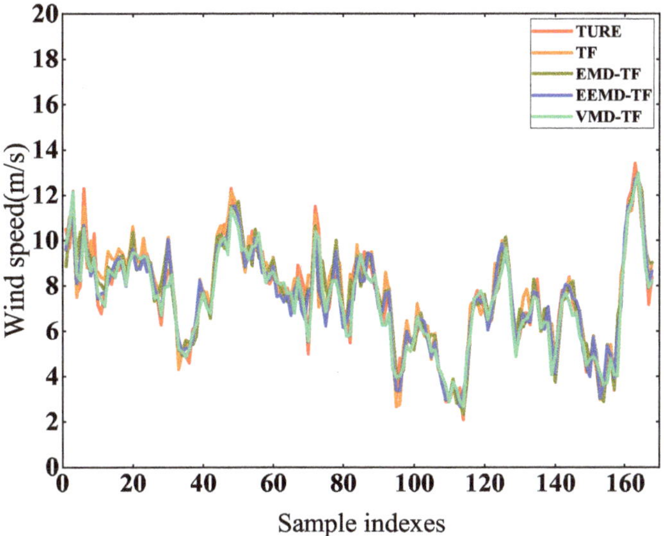

Figure 12. Comparison of the forecast and true wind speed for summer testing data.

According to Figure 12, even when the wind speed changes greatly VMD-TF is able to track and forecast well, while TF, EMD-TF, and EEMD-TF cannot respond as well to such mutations. According to Table 3, the forecasting result with VMD-TF is the best, while TF is the worst. Thus, we are able to conclude that signal decomposition methods can greatly enhance wind speed forecasting accuracy, and that of the methods investigated here, VMD shows the best performance.

Table 3. The forecasting errors of each model with the summer testing data.

Model	MAE	RMSE
TF	0.67	0.89
EMD-TF	0.56	0.78
EEMD-TF	0.47	0.65
VMD-TF	0.33	0.44

5.3.4. Effectiveness of M-Transformer

In order to illustrate that the M-Transformer model has good forecasting ability, we selected ARIMA, BP, the deep belief network (DBN), and LSTM as comparisons. These models, each composed of VMD and a single model, are referred to as VMD-ARIMA, VMD-BP, VMD-DBN, and VMD-LSTM, respectively. To assess the performance of these combined models, the winter testing set was used. Figure 13 compares the forecasting values and true values, while Table 4 shows the forecasting errors for each combined model.

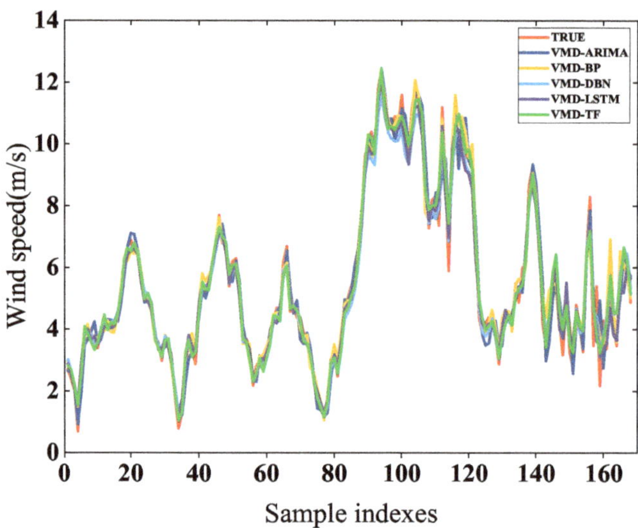

Figure 13. Comparison of forecast and real wind speed with winter testing data.

In Figure 13, all forecasting wind speed curves appear to be relatively close to the true wind speed curve. According to Table 4, however, the MAE and RMSE of VMD-TF are the smallest. Taking MAE as an example, the accuracy of VMD-TF decreased by 33%, 17%, 15%, and 9%, respectively, compared with the other four models, which shows that M-Transformer has superior performance.

Table 4. The forecasting errors of each combined model with the winter testing data.

Model	MAE	RMSE
VMD-ARIMA	0.43	0.59
VMD-BP	0.35	0.47
VMD-DBN	0.34	0.44
VMD-LSTM	0.32	0.42
VMD-TF	0.29	0.40

6. Conclusions

In this paper, we have proposed a variable support segment-based short-term wind speed forecasting model. Several conclusions can be drawn based on our experiments and analysis.

(1) VMD has a better decomposition effect than EMD and EEMD, and can effectively reduce the effects of wind speed non-stationarity.
(2) The M-Transformer model fully utilizes the characteristics of the self-attention mechanism, which can deeply mine potential information from wind speed series, estimate the variable support segment, and outperform other models in time series forecasting.
(3) VMD-TF combines the advantages of VMD and the self-attention mechanism, achieving significantly improved performance.

Although VMD-TF shows significant performance achievements, it neglects the impact of meteorological factors, which limits its ability to deal with sudden changes in wind speed. In future work, we intend to develop a model that is able to take into account both historical wind speed data and prevailing meteorological factors that influence wind speed.

Author Contributions: Conceptualization, K.Z.; methodology, K.Z. and X.L.; software, K.Z.; validation, X.L.; formal analysis, J.S.; investigation, X.L.; resources, J.S.; data curation, K.Z.; writing—original draft preparation, K.Z.; writing—review and editing, X.L. and J.S.; visualization, X.L.; supervision, J.S.; project administration, J.S. All authors have read and agreed to the published version of the manuscript.

Funding: This research received no external funding.

Institutional Review Board Statement: Not applicable.

Informed Consent Statement: Not applicable.

Data Availability Statement: Not applicable.

Conflicts of Interest: The authors declare no conflict of interest.

References

1. Shang, Z.; He, Z.; Chen, Y.; Xu, M. Short-term wind speed forecasting system based on multivariate time series and multi-objective optimization. *Energy* **2022**, *238*, 122024. [CrossRef]
2. Wang, H.Z.; Wang, G.B.; Li, G.Q.; Peng, J.C.; Liu, Y.T. Deep belief network based deterministic and probabilistic wind speed forecasting approach. *Appl. Energy* **2016**, *182*, 80–93. [CrossRef]
3. Dhiman, H.S.; Deb, D. A review of wind speed and wind power forecasting techniques. *arXiv* **2020**, arXiv:2009.02279.
4. Huang, X.; Wang, J.; Huang, B. Two novel hybrid linear and nonlinear models for wind speed forecasting. *Energy Convers. Manag.* **2021**, *238*, 114162. [CrossRef]
5. Liu, F.; Li, R.; Li, Y.; Cao, Y.; Panasetsky, D.; Sidorov, D. Short-term wind power forecasting based on TS fuzzy model. In Proceedings of the 2016 IEEE PES Asia-Pacific Power and Energy Engineering Conference (APPEEC), Xi'an, China, 5–28 October 2016; pp. 414–418.
6. Moreno, S.R.; Mariani, V.C.; Dos Santos Coelho, L. Hybrid multi-stage decomposition with parametric model applied to wind speed forecasting in Brazilian Northeast. *Renew. Energy* **2021**, *164*, 1508–1526. [CrossRef]
7. Fu, W.; Zhang, K.; Wang, K.; Wen, B.; Fang, P.; Zou, F. A hybrid approach for multi-step wind speed forecasting based on two-layer decomposition, improved hybrid DE-HHO optimization and KELM. *Renew. Energy* **2021**, *164*, 211–229. [CrossRef]
8. Sun, S.; Fu, J.; Li, A.; Zhang, P. A new compound wind speed forecasting structure combining multi-kernel LSSVM with two-stage decomposition technique. *Soft Comput.* **2021**, *25*, 1479–1500. [CrossRef]
9. Luo, L.; Li, H.; Wang, J.; Hu, J. Design of a combined wind speed forecasting system based on decomposition-ensemble and multi-objective optimization approach. *Appl. Math. Model.* **2021**, *89*, 49–72. [CrossRef]
10. Ren, C.; An, N.; Wang, J.; Li, L.; Hu, B.; Shang, D. Optimal parameters selection for BP neural network based on particle swarm optimization: A case study of wind speed forecasting. *Knowl. Based Syst.* **2014**, *56*, 226–239. [CrossRef]
11. Erdem, E.; Shi, J. ARMA based approaches for forecasting the tuple of wind speed and direction. *Appl. Energy* **2011**, *88*, 1405–1514. [CrossRef]
12. Santamaría-Bonfil, G.; Reyes-Ballesteros, A.; Gershenson, C. Wind speed forecasting for wind farms: A method based on support vector regression. *Renew. Energy* **2016**, *85*, 790–809. [CrossRef]
13. Nie, Y.; Liang, N.; Wang, J. Ultra-short-term wind-speed bi-forecasting system via artificial intelligence and a double-forecasting scheme. *Appl. Energy* **2021**, *301*, 117452. [CrossRef]
14. Zhou, Q.; Wang, C.; Zhang, G. A combined forecasting system based on modified multi-objective optimization and sub-model selection strategy for short-term wind speed. *Appl. Soft Comput.* **2020**, *94*, 106463. [CrossRef]
15. Li, H.; Wang, J.; Lu, H.; Guo, Z. Research and application of a combined model based on variable weight for short term wind speed forecasting. *Renew. Energy* **2018**, *116*, 669–684. [CrossRef]
16. Jiang, P.; Li, C. Research and application of an innovative combined model based on a modified optimization algorithm for wind speed forecasting. *Measurement* **2018**, *124*, 395–412. [CrossRef]
17. Wang, J.; Heng, J.; Xiao, L.; Wang, C. Research and application of a combined model based on multi-objective optimization for multi-step ahead wind speed forecasting. *Energy* **2017**, *125*, 591–613. [CrossRef]
18. Mallat, S.G. A theory for multiresolution signal decomposition: The wavelet representation. *IEEE Trans. Pattern Anal. Mach. Intell.* **1989**, *11*, 674–693. [CrossRef]
19. Huang, N.E.; Shen, Z.; Long, S.R.; Wu, M.C.; Shih, H.H.; Zheng, Q.; Yen, N.C.; Tung, C.C.; Liu, H.H. The empirical mode decomposition and the Hilbert spectrum for nonlinear and non-stationary time series analysis. *Proc. R. Soc. Lond. Ser. A Math. Phys. Eng. Sci.* **1998**, *454*, 903–995. [CrossRef]
20. Dragomiretskiy, K.; Zosso, D. Variational mode decomposition. *IEEE Trans. Signal Process.* **2013**, *62*, 531–544. [CrossRef]

21. Memarzadeh, G.; Keynia, F. A new short-term wind speed forecasting method based on fine-tuned LSTM neural network and optimal input sets. *Energy Convers. Manag.* **2020**, *213*, 112824. [CrossRef]
22. Naik, J.; Satapathy, P.; Dash, P.K. Short-term wind speed and wind power prediction using hybrid empirical mode decomposition and kernel ridge regression. *Appl. Soft Comput.* **2018**, *70*, 1167–1188. [CrossRef]
23. Moreno, S.R.; da Silva, R.G.; Mariani, V.C.; dos Santos Coelho, L. Multi-step wind speed forecasting based on hybrid multi-stage decomposition model and long short-term memory neural network. *Energy Convers. Manag.* **2020**, *213*, 112869. [CrossRef]
24. Jiang, P.; Liu, Z.; Niu, X.; Zhang, L. A combined forecasting system based on statistical method, artificial neural networks, and deep learning methods for short-term wind speed forecasting. *Energy* **2021**, *217*, 119361. [CrossRef]
25. Liu, H.; Mi, X.; Li, Y. Wind speed forecasting method based on deep learning strategy using empirical wavelet transform, long short term memory neural network and Elman neural network. *Energy Convers. Manag.* **2018**, *156*, 498–514. [CrossRef]
26. Altan, A.; Karasu, S.; Zio, E. A new hybrid model for wind speed forecasting combining long short-term memory neural network, decomposition methods and grey wolf optimizer. *Appl. Soft Comput.* **2021**, *100*, 106996. [CrossRef]
27. Neshat, M.; Nezhad, M.M.; Abbasnejad, E.; Mirjalili, S.; Tjernberg, L.B.; Garcia, D.A.; Alexander, B.; Wagner, M. A deep learning-based evolutionary model for short-term wind speed forecasting: A case study of the Lillgrund offshore wind farm. *Energy Convers. Manag.* **2021**, *236*, 114002. [CrossRef]
28. Hu, H.; Wang, L.; Tao, R. Wind speed forecasting based on variational mode decomposition and improved echo state network. *Renew. Energy* **2021**, *164*, 729–751. [CrossRef]
29. Khodayar, M.; Wang, J. Spatio-temporal graph deep neural network for short-term wind speed forecasting. *IEEE Trans. Sustain. Energy* **2018**, *10*, 670–681. [CrossRef]
30. Vaswani, A.; Shazeer, N.; Parmar, N.; Uszkoreit, J.; Jones, L.; Gomez, A.N.; Kaiser, Ł.; Polosukhin, I. Attention is all you need. *Adv. Neural Inf. Process. Syst.* **2017**, *30*, 5998–6008.
31. Santhosh, M.; Venkaiah, C.; Kumar, D.M.V. Ensemble empirical mode decomposition based adaptive wavelet neural network method for wind speed prediction. *Energy Convers. Manag.* **2018**, *168*, 482–493. [CrossRef]
32. Liu, H.; Mi, X.; Li, Y. An experimental investigation of three new hybrid wind speed forecasting models using multi-decomposing strategy and ELM algorithm. *Renew. Energy* **2018**, *123*, 694–705. [CrossRef]
33. Singh, S.N.; Mohapatra, A. Repeated wavelet transform based ARIMA model for very short-term wind speed forecasting. *Renew. Energy* **2019**, *136*, 758–768.

Article

Intelligent Modeling of the Incineration Process in Waste Incineration Power Plant Based on Deep Learning

Lianhong Chen [1], Chao Wang [1], Rigang Zhong [1], Jin Wang [2] and Zheng Zhao [2,*]

1 Shenzhen Energy Environment Engineering Co., Ltd., Shenzhen 518048, China; chenlianhong@sec.com.cn (L.C.); wangchaob@sec.com.cn (C.W.); zhongrigang@sec.com.cn (R.Z.)
2 School of Control and Computer Engineering, North China Electric Power University, Baoding 071003, China; wj07401347@163.com
* Correspondence: zheng_zhao@ncepu.edu.cn; Tel.: +86-158-0312-8060

Abstract: The incineration process in waste-to-energy plants is characterized by high levels of inertia, large delays, strong coupling, and nonlinearity, which makes accurate modeling difficult. Therefore, an intelligent modeling method for the incineration process in waste-to-energy plants based on deep learning is proposed. First, the output variables were selected from the three aspects of safety, stability and economy. The initial variables related to the output variables were determined by mechanism analysis and the input variables were finally determined by removing invalid and redundant variables through the Lasso algorithm. Secondly, each delay time was calculated, and a multi-input and multi-output model was established on the basis of deep learning. Finally, the deep learning model was compared and verified with traditional models, including LSSVM, CNN, and LSTM. The simulation results show that the intelligent model of the incineration process in the waste-to-energy plant based on deep learning is more accurate and effective than the traditional LSSVM, CNN and LSTM models.

Keywords: waste-to-energy; deep learning; variable selection; intelligent modeling

Citation: Chen, L.; Wang, C.; Zhong, R.; Wang, J.; Zhao, Z. Intelligent Modeling of the Incineration Process in Waste Incineration Power Plant Based on Deep Learning. *Energies* **2022**, *15*, 4285. https://doi.org/10.3390/en15124285

Academic Editor: Surender Reddy Salkuti

Received: 1 May 2022
Accepted: 10 June 2022
Published: 10 June 2022

1. Introduction

At present, the treatment methods for domestic waste usually include landfill, compost and incineration [1,2]. According to the statistics and the volume of domestic waste removal and transportation, the proportion of landfill treatment, compost treatment and incineration treatment is 58.30%, 2.10% and 36.20% respectively, and the remaining 3.40% are treated by simple landfill and stacking [3]. However, due to problems such as the large amount of land required and environmental pollution, the proportion of landfill treatment and compost treatment is decreasing year by year. The waste incineration process reduces the content of harmful substances in the waste by pyrolysis and oxidation under high temperature and high pressure. The volume of waste after incineration is reduced by more than 85% and the weight is reduced by more than 75%. The waste incineration process greatly eliminates the germs and harmful components in the waste, thus achieving the efficient treatment of the waste. Additionally, the energy generated by incineration can be used to generate electricity to realize a major goal of waste recycling [4]. It has been suggested that waste incineration power generation technology has the advantages of "reduction, recycling, and harmlessness", and that it is currently the best way to deal with domestic waste [5]. However, due to the complex composition of waste, the large fluctuations in waste calorific value, and the fact that the incinerator is a multi-input and multi-output (MIMO) object distinguished by high levels of inertia, large delays, strong coupling, and nonlinearity, it is difficult to meet the needs of the subsequent combustion optimization. Therefore, establishing an accurate and reliable intelligent model of the incineration process of waste-to-energy plants is the key to subsequent incineration optimization [6,7].

Elisa [8] et al. used the mechanism modeling method to model the incineration process of waste-to-energy power plants, but there were problems related to the complicated derivation process and low precision. Therefore, data-driven modeling has been widely used in combustion process modeling [9,10]. Peng et al. [11] established a multi-input and single-output model for boiler combustion oxygen content based on big data and a neural network, and enhanced the neural network through Bayesian arithmetic, which solved the problem of slow learning speed and the problem of obtaining the optimal value in a small range of the classical neural network. Based on the operating data for a boiler in a thermal power plant, Song et al. [12] used a radial basis neural network to establish a model with the flue gas oxygen content, furnace negative pressure and steam pressure as outputs. Compared with a back propagation (BP) neural network, a radial basis function (RBF) neural network has better categorization capability, approximation ability and learning speed, but it has poor resistance to noise in the sample data. Zhong et al. [13] used the particle swarm algorithm and support vector machine to establish a boiler exhaust gas temperature model of a 660 MW unit, which provided guidance for the operation of the boiler. However, for large amounts of sample data, support vector machines are prone to overfitting and lack modeling accuracy. Due to the structural limitations of the algorithms themselves, these algorithms cannot mine the deep information in the sample data [14].

With the rapid development of artificial intelligence, modeling methods based on deep learning have attracted more and more scholars' attention. Hu et al. [15] established a boiler combustion efficiency model using a convolutional neural network for a 600 MW supercritical unit boiler in Henan. Yu et al. [16] used deep CNN and support vector machine to extract and analyze the deep features of flame images, and realized the modeling of the NOx concentration of a 4.2 MW heavy oil combustion boiler. Zhang [17] established a deep neural network model of a stacking noise reduction autoencoder and LSTM network considering the characteristics of ultra-supercritical units such as high inertia, large delays and noise in the actual data. However, the existing modeling methods generally have defects such as too few input and output variables, which are far from the actual operation of the actual unit, and the inability to express the dynamic characteristics of the model. Therefore, in the process of model establishment, in addition to selecting the modeling method, the selection of input variables will also affect the modeling accuracy. Wang et al. [18] utilized principal component analysis means to lower the dimension of the multi-dimensional input variables of wind turbines. Although the feature dimension was reduced, the original data was changed and the interpretability of the model was reduced.

Incinerator incineration process modeling data are characterized as complex large sample data, nonlinear time series, etc. Compared with the methods described above, the deep learning network method can use the complexity relationship between data to automatically model and adjust the model parameters so as to establish the optimal nonlinear model between input and output. That is, this method is able to use the time series' characteristics or other complex relationships between historical data to model through deep learning networks. In summary, an intelligent model of the incineration process of waste-to-energy plants based on deep learning is elicited. First, the output variables are selected from the three aspects of safety, stability and economy. The initial variables related to the output variables are determined. The input variables were finally determined by removing the invalid variables and redundant variables through the Lasso algorithm. Secondly, each delay time is calculated, and a multi-input and multi-output model based on deep learning is established. Finally, the model proposed in this paper is compared with the traditional model to verify the improvement in its accuracy.

2. Basic Method Principle

2.1. Waste-to-Energy Treatment Technology

After the garbage is transported to the incineration plant, it is fermented in a garbage storage tank for 3–5 days to increase the calorific value. The calorific value of the waste after fermentation is about 1800–2100 kcal/kg. Garbage fermentation mainly relies on the role of microorganisms in the garbage. At the same time, during the storage of the garbage, the water in the garbage is continuously leached out. After storage and fermentation, the garbage is moved to the feeding hopper by a hanging garbage grab, and is transported to the incinerator through the feeding grate. The incinerator is composed of a multi-stage mechanical grate, of which the first and second stages of the incineration grate are the waste drying area, the third and fourth stages of the incineration grate are the waste gasification area (main combustion area), and the fifth stage is the burning ember area. At the same time, each section of the combustion grate includes a fixed grate, sliding grate and turning grate. The sliding grates slowly push the garbage layer forward on the grate, and the function of the turning grate is to drive the overturning grate pieces to turn up and down through the reciprocating rotation of the overturning shaft, so as to support the garbage bed in a local position, destroy the original garbage bed, cause the previously formed bed to dislocate, and break the hard-shell surface and molten layer caused by burning. The primary combustion air is blown into the interior of the garbage bed (like throwing a fire) so that the garbage can be completely burned. The fan system consists of a primary air system, secondary air system and furnace-wall-cooling air system, among the first two wind systems that affect the combustion process. The primary air is extracted from the garbage storage tank, heated to about 180 °C by the air preheater, and sent to the bottom of the incinerator through the gap between the grate pieces of the incinerator. Then it penetrates the garbage bed and enters the incinerator chamber, where it is mixed with the garbage and burns. At the same time, a negative pressure in the garbage pond is established to prevent the overflow of garbage odor, so as to achieve effective management of odor. The secondary air mainly adjusts the amount of oxygen to ensure better combustion conditions. In view of the characteristics of China's garbage, which has high moisture content, high non-combustible content and low calorific value, the design of the rear arch of the furnace is adapted, as shown in Figure 1, to form a good aerodynamic field and help combustion [19].

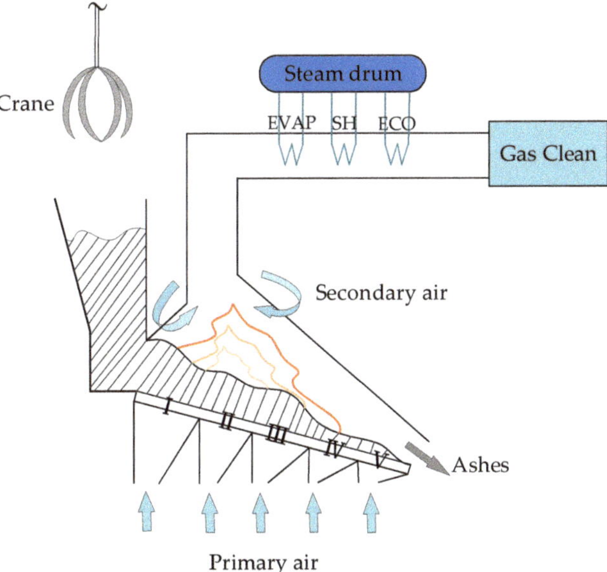

Figure 1. Incinerator structure diagram.

2.2. Lasso Algorithm

Least absolute shrinkage and selection operator (Lasso) is a penalty-based variable selection method first proposed by the famous statistician Robert Tibshirani in 1996 [20,21]. The specific principle is as follows.

The following is an example of a typical linear regression model:

$$y_i = \beta_0 + \sum_{j=1}^{p} x_{ij}\beta_j + e_i \quad i = 1, 2, \cdots, n \tag{1}$$

There are n sets of observations, and each group of observations consists of an input variable y_i and p-correlated predictor variables $x_i = (x_{i1}, x_{i2}, \cdots, x_{ip})^T$.

The traditional method is to minimize the least squares objective function:

$$\underset{\beta_0,\beta}{\text{min}mize} \sum_{i=1}^{n} (y_i - \beta_0 - \sum_{j=1}^{p} x_{ij}\beta_j)^2 \tag{2}$$

In the general formula, the least squares estimation of β is not 0, and if $n < p$, there are countless solutions that make the objective function 0, hence the result of the least squares estimation is not unique. Therefore, this process needs to introduce a penalty function, that is, regularization. The Lasso algorithm is based on the least squares estimation to introduce a penalty factor to constrain the norm of β, as shown in the formula.

$$\begin{gathered} RSS(\hat{\beta}) = \underbrace{\text{argmin}}_{\hat{\beta}} \sum_{i=1}^{n} (y_i - \beta_0 - \sum_{j=1}^{p} \beta_j x_{ij})^2 + \lambda \sum_{j=1}^{p} |\beta_j| \\ \hat{\beta} = \text{argmin}\left\{ \|y - \beta_0 - x\beta\|_2^2 \right\} \\ s.t \quad \|\beta\| \leq t \end{gathered} \tag{3}$$

In the formula, $\lambda \geq 0$ is the hyperparameter, $\lambda \sum_{j=1}^{p} |\beta_j|$ is the compression penalty, t is the adjustment parameter, and the inequality $\|\beta\| \leq t$ effectively restricts the parameter space and realizes feature selection.

2.3. Model Building Based on Deep Learning

2.3.1. Convolutional Neural Network

CNN is a feedforward network that was first used in image processing and has excellent performance [22]. CNN has the characteristics of weight sharing, local connection, and dimensionality reduction sampling, and can fully mine the local characteristics of the data itself.

CNN generally contains three basic layers: a convolutional layer, pooling layer and fully connected layer. The pixels in the local area of the input image are weighted by the weight coefficient of the convolution kernel, the operation of feature extraction is completed by the convolution layer, and the activation function introduces nonlinear changes to the network model. The pooling layer performs dimension reduction sampling on the output of the convolutional layer, and at the same time, the pooling operation results in translation invariance in the CNN. The fully connected layer is where each node is linked to all the nodes in the previous layer, which is used to synthesize the features extracted in the front, as depicted in Figure 2.

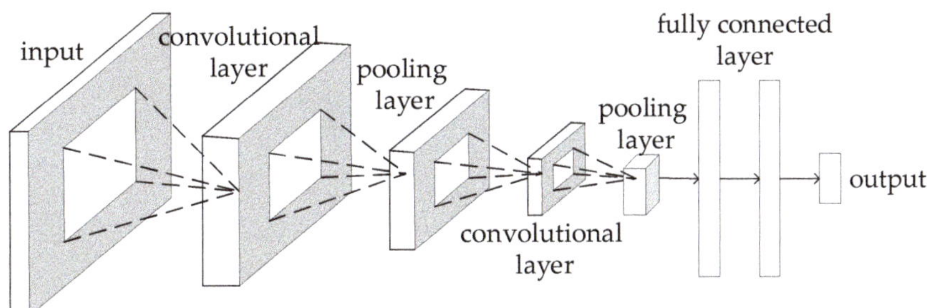

Figure 2. Construction of CNN.

This paper uses CNN to fully mine the features of the data, and the feature data processed by the convolution operation is sent to the Bi-LSTM network for further operations.

2.3.2. Bi-LSTM Model

The cyclic unit construction of LSTM is exhibited in Figure 3.

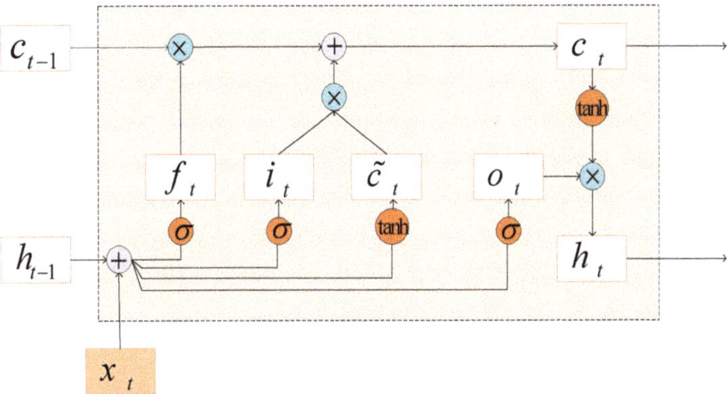

Figure 3. Cyclic unit construction of the LSTM network.

$$\begin{bmatrix} \widetilde{c}_t \\ o_t \\ i_t \\ f_t \end{bmatrix} = \begin{bmatrix} \tanh \\ \sigma \\ \sigma \\ \sigma \end{bmatrix} \left(P \begin{bmatrix} x_t \\ h_{t-1} \end{bmatrix} + b \right) \tag{4}$$

$$i_t = \sigma(P_i x_t + Q_i h_{t-1} + b_i) \tag{5}$$

$$f_t = \sigma\left(P_f x_t + Q_f h_{t-1} + b_f\right) \tag{6}$$

$$o_t = \sigma(P_O x_t + Q_o h_{t-1} + b_O) \tag{7}$$

$$c_t = f_t \otimes c_{t-1} + i_t \otimes \widetilde{c}_t \tag{8}$$

$$\widetilde{c}_t = \tanh(P_c x_t + Q_c h_{t-1} + b_c) \tag{9}$$

$$h_t = o_t \otimes \tanh(c_t) \tag{10}$$

First, use the input x_t at the present moment and the external condition h_{t-1} at the previous moment to calculate f_t, i_t, o_t and $\widetilde{c}_t$. Secondly, use f_t and i_t to update the memory unit c_t, and finally, pass the internal state information to the external state h_t in combination with o_t [23].

LSTM can only extract forward sequence information, while Bi-LSTM (Bidirectional Long Short-Term Memory) extracts sequence information in both directions to obtain more data features [24]. Bi-LSTM consists of two layers of LSTM networks with the same input and different information transfer directions. The Bi-LSTM structure is shown in Figure 4.

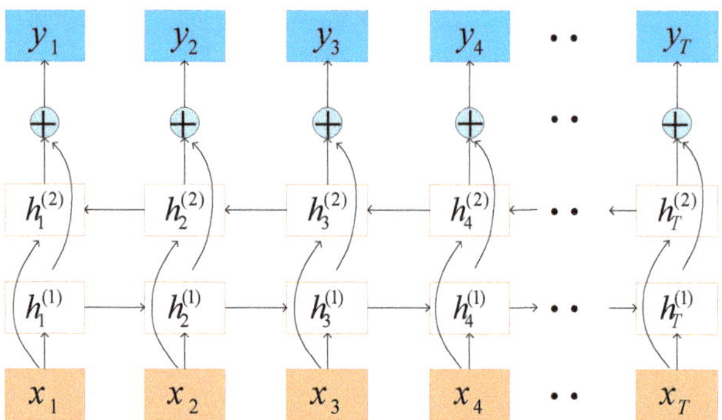

Figure 4. Two-way cyclic neural network expanded by time.

$$h_t^{(1)} = f(U^{(1)}h_{t-1}^{(1)} + W^{(1)}x_t + b^{(1)}) \tag{11}$$

$$h_t^{(2)} = f(U^{(2)}h_{t+1}^{(2)} + W^{(2)}x_t + b^{(2)}) \tag{12}$$

$$h_t = h_t^{(1)} \oplus h_t^{(2)} \tag{13}$$

$\oplus$ is the vector concatenation.

2.3.3. Intelligent Model of Incineration Process Based on Deep Learning

The CNN-BiLSTM combined model not only combines the feedforward mechanism of the CNN with the feedback mechanism of the RNN, but also greatly reduces the computational cost through feature extraction of the CNN. Furthermore, by combining the models with BiLSTM, the model accuracy is improved.

Unlike steady-state modeling, dynamic models take into account the effects of time [25]. In the dynamic model of the CNN-BiLSTM incineration process, the output is not only related to the current data of each auxiliary variable, but also related to the delay time of the input and output variables [26].

In describing the dynamic characteristics of the waste incinerator combustion process, due to the existence of the delay time d, the sampling point at time t can be represented as $\{x(t-d), y(t)\}$. By discretizing the dynamic model of the incineration process, the difference equation form of the dynamic model is obtained as:

$$y(t) = f[y(t-1), \cdots y(t-n), x(t-d)] \tag{14}$$

The above formula can be expressed as an intelligent model of the incineration process of waste-to-energy plants, which is a MIMO model. From the above equation, the output variable of the model can be obtained and is shown as the relationship of the output values of n past moments and the input values of d past moments.

3. Intelligent Model of Waste-to-Energy Plant Incineration Process Based on Deep Learning

3.1. *Variable Selection Based on the Lasso Algorithm*

Before building an incineration model, the input variables and output variables of the model should be determined. In this paper, the selection of the output variables of the intelligent model of waste-to-energy plant incineration process took the three aspects of safety, stability and economy into consideration. The output variables are the oxygen content of the boiler flue gas outlet (C_{O_2}), steam flow (Q), the furnace temperature of the incinerator (T_1), and the temperature of the ember section (T_2). The amount of oxygen is related to the load, and the amount of oxygen is used as a precursor to load changes. When the oxygen feedback value is higher than the set value, it means that the air volume is excessive or the garbage calorific value is insufficient. At this time, the boiler load decreases. When the oxygen feedback value is lower than the set value, the boiler load will increase. At this time, the boiler load should be reduced. If the oxygen content is lower than a certain level, it means that there is an abnormal situation such as an explosion in the furnace. At this time, the feeding should be stopped to prevent the garbage in the furnace from deflagrating and causing danger. When the steam flow is stable, this ensures that the steam turbine and generator work at the rated load and the equipment performance is good. Ensuring that the furnace flue gas stays above 850 °C for 2 s can prevent the generation of harmful flue gas dioxins. The temperature of the ember section is maintained within a certain range, which ensures that the garbage is fully burned and improves the combustion efficiency.

Mechanism analysis was used to select the input variables that have an impact on the output variables. A total of 19 variables related to the output variables were screened out from the variables collected by the power plant, namely, primary air flow, the temperature of primary air, unit 1–5 primary air flow, secondary air flow, the temperature of secondary air, unit 1–5 material layer thickness, the transmission speed of the sliding grate unit 1, the transmission speed of the sliding grate unit 2, the transmission speed of the sliding grate unit 3, the transmission speed of the sliding grate unit 4, and the transmission speed of the sliding grate unit 5. Since the secondary air temperature basically does not fluctuate, and there is a coefficient relationship between the transmission speeds of the five units of the sliding grate, 19 variables were screened and 14 initial variables were obtained.

The 14 initial variables were selected by the Lasso algorithm, and 8 input variables were finally obtained, namely, primary air flow, unit 1 primary air flow, unit 4 primary air flow, unit 5 primary air flow, secondary air flow, unit 1 material layer thickness, material layer thickness of unit 5, and conveying speed of sliding grate unit 1. Then, 25,200 sets of data were selected from a northern waste power plant from 16:00 on 6 August 2019 to 10:00 on 8 August 2019, and the sampling time was 6 s. The unit and variation range of each input variable are shown in Table 1. The local trend diagrams of the input variables are shown in Figure 5.

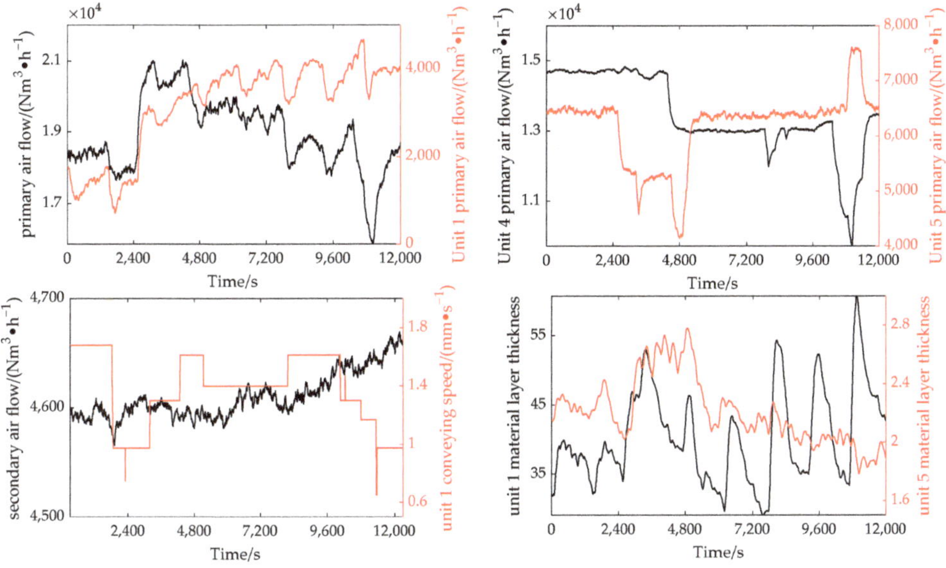

Figure 5. The local trend diagrams of the input variables.

Table 1. Unit and variation range of input variables.

Serial Number	Variable Name	Unit	Variation Range
1	primary air flow	Nm^3/h	13,500–23,069
2	Unit 1 primary air flow	Nm^3/h	299–5932
3	Unit 4 primary air flow	Nm^3/h	9721–17,763
4	Unit 5 primary air flow	Nm^3/h	1423–13,173
5	secondary air flow	Nm^3/h	4538–4673
6	unit 1 material layer thickness	-	11.98–74.25
7	unit 5 material layer thickness	-	1.27–5.05
8	unit 1 conveying speed of sliding grate	mm/s	0.14–2.80

3.2. *Calculation of Delay Time*

Waste-to-energy generating units typically have large delays, and there is a time delay between the various data collected by the power plant. When an input variable changes, it takes a while for the output variable to react to the change. In order to ensure the consistency of each input variable and output variable in the time sequence, a time delay compensation algorithm based on mutual information is proposed. Mutual information can calculate the correlation between the two groups of samples. By determining the mutual information numerical value between the input variables and the output variables, the delay time between the input variables and the output variables can be obtained (the specific process can be found in [26]). Taking the steam flow as the output as an example, Table 2 shows the maximum mutual information of each input variable for the steam flow and the corresponding time delay at this time.

Table 2. Delay time and maximum mutual information of auxiliary variables.

Auxiliary Variable Number	1	2	3	4	5	6	7	8
Delay Time	260	290	240	90	210	280	70	250
Maximum Mutual Information	0.7794	0.6983	0.8434	1.0875	0.9343	0.7260	1.2867	0.7999

3.3. *Model Establishment*

The steps to build an intelligent model of the incineration process of a waste-to-energy plant are as follows:

(1) The initial variables are screened by mechanism analysis and the Lasso algorithm, and invalid variables and redundant variables are removed.
(2) Data preprocessing, including outlier removal, noise reduction, and normalization.
(3) The mutual information method is used to determine each delay time.
(4) The model is established: first, the input variables after feature selection are input into the CNN layer of the model, and the deep time series features are extracted through the convolution and pooling layers. Secondly, they are sent to the BiLSTM layer to further strengthen the connection between the temporal features. The last layer is the fully connected layer, and the model output is completed.

The CNN-BiLSTM model structure includes: 1 CNN, 1 max pooling layer, 1 dropout layer, 2 BiLSTM layers, and 1 fully connected layer. The parameters of the final training model are: the batch size is 128, epochs are 50, and the Adam optimization algorithm is selected.

In Figure 6, u is the initial variable, x is the input variable after variable selection, d_{ij} is the delay time between the input x_i and the output y_j, $\mathbf{Y}$ is the actual value, and $\hat{\mathbf{Y}}$ is the model output value.

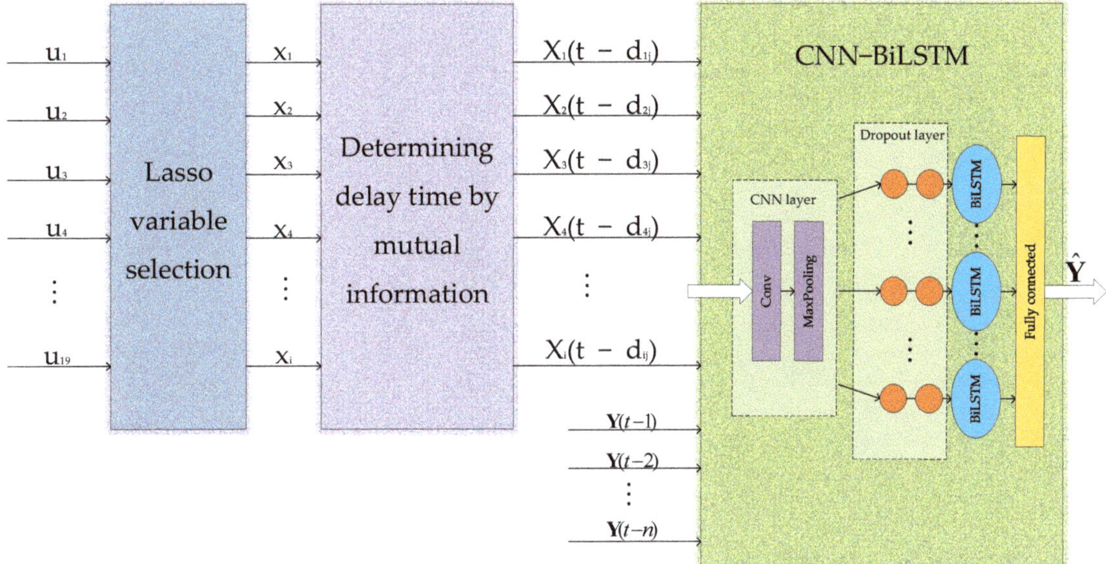

Figure 6. Modeling process diagram.

The model building process is shown in Figure 6.

4. Model Establishment and Result Analysis

4.1. Model Evaluation Indicators

The model evaluation indicators used in this paper are MAE, MAPE, and RMSE. The closer the value is to 0, the more accurate the output of the model. The calculation formula is as follows:

$$MAE = \frac{1}{n}\sum_{i=1}^{n}|y_i - \hat{y}_i| \tag{15}$$

$$MAPE = \frac{1}{n}\sum_{i=1}^{n}\left|\frac{y_i - \hat{y}_i}{y_i}\right| \times 100\% \tag{16}$$

$$RMSE = \sqrt{\frac{\sum_{i=1}^{n}(y_i - \hat{y}_i)^2}{n}} \tag{17}$$

where y_i is the actual sample value, $\hat{y}_i$ is the model output value, and n is the amount of data.

MAE reflects the magnitude of the deviation of the measured value from the true value; MAPE reflects the degree to which the sample output value deviates from the measured value; and RMSE reflects the sample standard deviation of the bias between the model output value and the measured value, reflecting the degree of dispersion of the sample. The combination of the three can better represent the precision of the model.

4.2. Model Establishment Result Analysis

.To test and verify that the output of the model is accurate, the first 90% of the 25,200 sets of data were used as the training set and the last 10% ere used as the test set in chronological order. Before the data is entered into the model, the data should be normalized to eliminate the influence of the input variables on the output of the modeling results due to different magnitudes and speeding up the running time of the model.

4.2.1. The Influence of Variable Selection on Modeling Results

There were still invalid variables and redundant variables in the variables obtained after the mechanism analysis. So as to solve the issue of model overfitting and further simplify the model, the Lasso algorithm was used for variable selection. Experiments were executed on the 14 initial variables obtained from the mechanism analysis and the 8 input variables selected by the Lasso algorithm to verify that the output of the model was accurate. The final evaluation indexes are revealed in Table 3.

Table 3. Evaluation index of the influence of variable selection on modeling results.

	Before Variable Selection (MAE/MAPE/RMSE)	After Variable Selection (MAE/MAPE/RMSE)
T_1 (°C)	3.348/0.293/4.299	3.245/0.284/4.027
Q (t/h)	0.060/0.232/0.075	0.051/0.196/0.064
CO_2 (%)	0.101/2.530/0.130	0.100/2.519/0.130
T_2 (°C)	0.274/0.121/0.407	0.240/0.100/0.364

It can be seen from Table 3 that after the selection of the initial variables, the test set error decreased. This phenomenon shows that since there are invalid variables and redundant variables in the initial variables, if these variables continue to be retained in the input variables, the generalization ability of the model will be reduced. Therefore, it is more efficient to select the variables before building a model as this not only simplifies the model and reduces the computing time during modeling, but also prevents the model from overfitting and improves the model accuracy.

4.2.2. The Influence of Different Models on the Modeling Results

Three classic models, LSSVM, CNN, and LSTM were selected for comparison to verify the effectiveness of the CNN-BiLSTM model. When other conditions were kept the same, the modeling accuracy is as shown in Table 4, and a comparison of the modeling results of the CNN-BiLSTM model is presented in Figure 7.

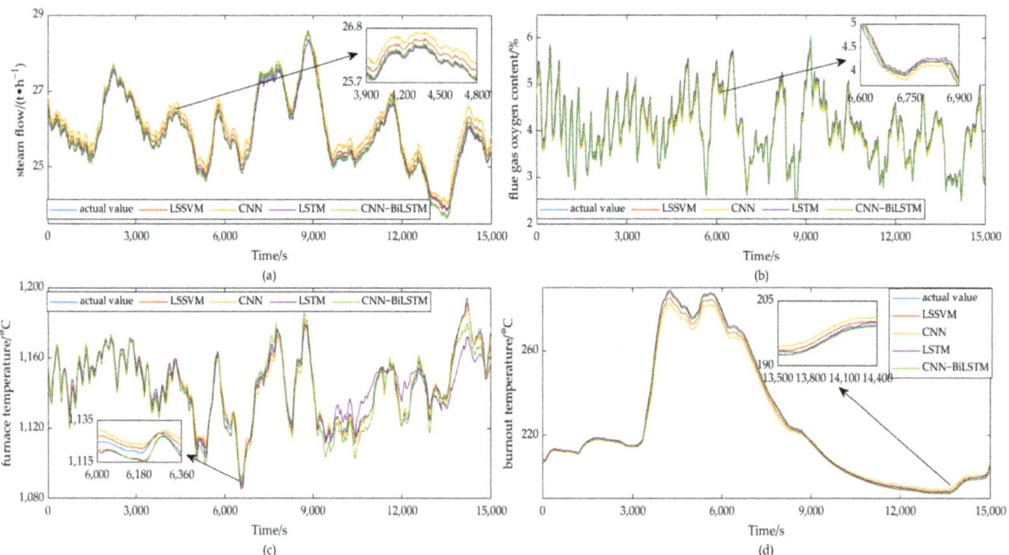

Figure 7. Modeling result diagram. (**a**) Comparison of output results of steam flow model; (**b**) comparison of the output results of the flue gas oxygen content model; (**c**) comparison of output results of the furnace temperature model; (**d**) comparison of the output results of the temperature model in the ember stage.

Table 4. Influence of different models on modeling accuracy.

	T_1 (°C) (MAE/MAPE/RMSE)	Q (t/h) (MAE/MAPE/RMSE)	C_{O_2} (%) (MAE/MAPE/RMSE)	T_2 (°C) (MAE/MAPE/RMSE)
LSSVM	4.226/0.423/5.239	0.178/0.654/0.432	0.156/3.106/0.177	1.324/0.849/1.637
CNN	4.540/0.395/6.941	0.292/1.140/0.316	0.133/3.290/0.159	2.198/0.920/2.800
LSTM	3.899/0.350/4.397	0.066/0.258/0.087	0.111/2.637/0.134	0.597/0.262/0.652
CNN-BiLSTM	3.245/0.284/4.027	0.051/0.196/0.064	0.100/2.519/0.130	0.240/0.100/0.364

Table 4 shows that, taking the steam flow as an example, the mean absolute error of LSSVM is 0.178, the mean absolute error of CNN is 0.292, the mean absolute error of LSTM is 0.066, and the mean absolute error of CNN-BiLSTM is 0.051. So, the order of model accuracy from low to high is CNN, LSSVM, LSTM, CNN-BiLSTM. It can be seen that the intelligent model based on deep learning can effectively improve the utility value of the traditional model, and it has stronger generalization ability and modeling accuracy.

5. Conclusions

A combined model based on feature selection and CNN-BiLSTM was constructed in this paper. First, the Lasso algorithm was used to remove invalid and redundant variables from the initial variables to determine the input variables, and the effective feature information was extracted through the CNN network. Finally, the BiLSTM network was used to train the model. Historical operation data from a waste-to-energy plant in the north was used for simulation analysis. The main conclusions are as follows:

(1) In this paper, based on the historical operation data from waste-to-energy power plants, multi-dimensional feature sets including waste factors, grate operation factors, and air volume factors were used, and high-correlation feature parameters through the effective feature screening of multi-dimensional feature sets by Lasso algorithm were selected. The comparison of before and after feature selection shows that Lasso feature screening for multi-dimensional input feature parameters can improve model accuracy.

(2) Compared with the traditional LSSVM, CNN, and LSTM models, the bidirectional network model based on feature selection and CNN-BiLSTM selected in this paper, can fully mine data features under multi-dimensional input feature parameters, and it has higher accuracy and applicability.

Author Contributions: Conceptualization, L.C., C.W. and R.Z.; methodology, C.W., R.Z. and J.W.; software, L.C., R.Z. and J.W.; validation, R.Z., C.W. and J.W.; formal analysis, L.C., C.W. and R.Z.; investigation, L.C., R.Z. and J.W.; resources C.W., J.W. and Z.Z.; data curation, L.C., C.W. and R.Z.; writing—original draft preparation, C.W., R.Z. and J.W.; writing—review and editing, L.C., C.W., R.Z., J.W. and Z.Z.; visualization, C.W., R.Z. and J.W.; supervision, L.C., C.W. and R.Z.; project administration, C.W., R.Z. and J.W.; funding acquisition, R.Z. All authors have read and agreed to the published version of the manuscript.

Funding: This research was funded by the Shenzhen Special Sustainable Development Science and Technology Project, grant number KCXFZ20201221173402007.

Institutional Review Board Statement: Not applicable.

Informed Consent Statement: Not applicable.

Data Availability Statement: Not applicable.

Conflicts of Interest: The authors declare no conflict of interest.

References

1. Chen, S.; Huang, J.; Xiao, T.; Gao, J.; Bai, J.; Luo, W.; Dong, B. Carbon emissions under different domestic waste treatment modes induced by garbage classification. *Sci. Total Environ.* **2020**, *717*, 137193. [CrossRef] [PubMed]
2. Moschou, C.E.; Papadimitriou, D.M.; Galliou, F.; Markakis, N.; Papastefanakis, N.; Daskalakis, G.; Sabathianakis, M.; Stathopoulou, E.; Bouki, C.; Daliakopoulos, I.N.; et al. Grocery Waste Compost as an Alternative Hydroponic Growing Medium. *Agronomy* **2022**, *12*, 789. [CrossRef]
3. Xu, H. A summary of the development of the municipal solid waste treatment industry in 2017. *China's Environ. Prot. Ind.* **2018**, *4*, 5–9.
4. Huang, M.; He, W.; Incecik, A.; Cichon, A.; Królczyk, G.; Li, Z. Renewable energy storage and sustainable design of hybrid energy powered ships: A case study. *J. Energy Storage* **2021**, *43*, 103266. [CrossRef]
5. Sun, G.; Tian, J.; Su, M. *The Foundation and Application of Domestic Waste Incineration Power Generation Technology*; Hefei University of Technology Press: Hefei, China, 2019.
6. Liu, J. *Combustion System Modeling and Optimization Based on Deep Learning*; Shanghai Jiaotong University: Shanghai, China, 2016.
7. Sun, Y.; Peng, G.; Jin, K.; Liu, S.; Gardoni, P.; Li, Z. Force/motion transmissibility analysis and parameters optimization of hybrid mechanisms with prescribed workspace. *Eng. Anal. Bound. Elem.* **2022**, *139*, 264–277. [CrossRef]
8. Magnanelli, E.; Tranås, O.L.; Carlsson, P.; Mosby, J.; Becidan, M. Dynamic modeling of municipal solid waste incineration. *Energy* **2020**, *209*, 118426. [CrossRef]
9. Zhang, Y.; Pan, G.; Zhao, Y.; Li, Q.; Wang, F. Short-term wind speed interval prediction based on artificial intelligence methods and error probability distribution. *Energy Convers. Manag.* **2020**, *224*, 113346. [CrossRef]
10. Zhang, Y.; Zhao, Y.; Shen, X.; Zhang, Z. A comprehensive wind speed prediction system based on Monte Carlo and artificial intelligence algorithms. *Appl. Energy* **2022**, *305*, 117815. [CrossRef]
11. Peng, D.; Mei, L.; Li, S.; He, J. Research of modeling for the oxygen content of boiler combustion based on large data and neural network. *Therm. Power Eng.* **2018**, *33*, 86–92.
12. Song, Q.; Li, Y. Modeling of the boiler combustion system by RBF neural networks. *J. Harbin Univ. Sci. Technol.* **2016**, *21*, 89–92.
13. Zhong, Y.; Wei, H.; Yao, W.; Chen, J. Boiler exhaust gas temperature modeling based on particle swarm algorithm and support vector machine. *Therm. Power Gener.* **2016**, *45*, 32–36.
14. Duan, Y.; Lv, Y.; Zhang, J.; Zhao, X. Deep learning for control: The state of the art and prospects. *Acta Autom. Sin.* **2016**, *42*, 643–654.
15. Hu, H.; Zhang, J.; Liu, H.; Li, M.; Yang, Q. Power plant boiler combustion efficiency modeling approach based on convolutional neural networks. *J. Xi'an Jiaotong Univ.* **2019**, *53*, 10–15.
16. Yu, Y.; Han, Z.; Xu, C. NOx concentration prediction based on deep convolution neural network and support vector machine. *Chin. Soc. Electr. Eng.* **2022**, *42*, 238–248.
17. Zhang, H. *Application of Deep Neural Network in Electric Power Industry Modeling*; North China Electric Power University (Beijing): Beijing, China, 2020.
18. Wang, Y.; Xie, D.; Wang, X.; Li, G.; Zhu, M. Prediction of interaction between grid and wind farms based on PCA-LSTM model. *Chin. Soc. Electr. Eng.* **2019**, *39*, 4070–4081.
19. Huang, M.; Borzoei, H.; Abdollahi, A.; Li, Z.; Karimipour, A. Effect of concentration and sedimentation on boiling heat transfer coefficient of GNPs-SiO_2/deionized water hybrid Nanofluid: An experimental investigation. *Int. Commun. Heat Mass Transf.* **2021**, *122*, 105141. [CrossRef]
20. Tibshirani, R.J. Regression Shrinkage and Selection via the LASSO. *J. Ro-Yal Stat. Soc. Ser. B Methodol.* **1996**, *73*, 273–282. [CrossRef]
21. Hou, D. *Comparative Study and Empirical Analysis of Lasso Type Variable Selection Methods*; Shan Dong University: Jinan, China, 2021.
22. Luo, Y.; Fan, Y.; Chen, X. Research on optimization of deep learning algorithm based on convolutional neural network. *J. Phys. Conf. Ser.* **2021**, *1848*, 012038. [CrossRef]
23. Burgess, J.; O'Kane, P.; Sezer, S.; Carlin, D. LSTM RNN: Detecting exploit kits using redirection chain sequences. *Cybersecurity* **2021**, *4*, 25. [CrossRef]
24. Charu, C. *Aggarwal. Neural Networks and Deep Learning*; Springer: Cham, Switzerland, 2018.
25. Shen, Q. *Study of Dynamic Modeling and Optimization Control Method for Boiler Combustion System*; Southeast University: Nanjing, China, 2017.
26. Zhao, Z.; Li, Y.; Yuan, H. Dynamic model for soft sensing of NOx generation in coal-fired units. *J. Power Eng.* **2020**, *40*, 523–529.

MDPI

Article

HOG-SVM-Based Image Feature Classification Method for Sound Recognition of Power Equipments

Kang Bai [1], Yong Zhou [2], Zhibo Cui [1,*], Weiwei Bao [2], Nan Zhang [2] and Yongjie Zhai [1]

[1] Department of Automation, North China Electric Power University, Baoding 071003, China; baikang_zdh@ncepu.edu.cn (K.B.); zhaiyongjie@ncepu.edu.cn (Y.Z.)

[2] SPIC Central Research Institute, Beijing 102209, China; zhouyong0615@163.com (Y.Z.); alndr@163.com (W.B.); zhangnan19900321@126.com (N.Z.)

* Correspondence: czb@ncepu.edu.cn

Abstract: In this paper, a method of power system equipment recognition based on image processing is proposed. Firstly, we carry out wavelet transform on the sound signal of power system equipment collected from the site, and obtain the wavelet coefficient–time diagram. Then, the similarity of wavelet coefficients–time images of different equipment and the same equipment in different periods is calculated, which is used as the basis of the feasibility of image recognition. Finally, we select the HOG features of the image, and classify the selected features using SVM classifier. The method proposed in this paper can accurately identify and classify power system equipment through sound signals, and is different from the traditional method of classifying sound signals directly. The advantages of image processing can be effectively utilized through image processing to avoid the limitations of sound signal processing.

Keywords: electric power equipment; voice recognition; HOG feature extraction; SVM classifier; image processing

Citation: Bai, K.; Zhou, Y.; Cui, Z.; Bao, W.; Zhang, N.; Zhai, Y. HOG-SVM-Based Image Feature Classification Method for Sound Recognition of Power Equipments. *Energies* **2022**, *15*, 4449. https://doi.org/10.3390/en15124449

Academic Editors: Abu-Siada Ahmed and Jong Hoon Kim

Received: 18 May 2022
Accepted: 16 June 2022
Published: 18 June 2022

Publisher's Note: MDPI stays neutral with regard to jurisdictional claims in published maps and institutional affiliations.

1. Introduction

With the gradual development of large-scale, integrated, highly automated and intelligent power system equipment, not only are rapid economic benefits introduced, but also the risk of great loss caused by sudden equipment failure is increased. Therefore, the comprehensive, timely and accurate monitoring of the power system equipment health status ensures the stable operation of equipment, reduces the accidental shutdown rate and has a high investment–income ratio. To this end, researchers carried out systematic research on temperature, vibration, image and other aspects of various power system equipment, and obtained effective information characteristics [1–3]. In addition, artificial intelligence [4], deep learning [5] and neural network [6] have been used to realize fault monitoring of equipment.

According to Kafeel et al. [7], current, sound and vibration are the most commonly monitored parameters. In Ribeiro et al. [8] a hydro-generator current-monitoring system is proposed and the fast Fourier transform (FFT) is applied to the Parker transform of the current. Song et al. [9] used the bin method, the method based on multivariate normal distribution and the Copula method to compare three Bayesian diagnosis models on account of SCADA (Supervisory Control And Data Acquisition). Li et al. [10], aiming at the problems of high-speed and long-distance transmission and greatly increasing data storage capacity, proposed a method on account of adjustable q-factor wavelet transform morphologic module analysis, including few and scattered Bayesian iterative arithmetic unite stepping pulse dictionary. Yu et al. [11] try to build a rough set with feature relationships, then use a distribution reduction arithmetic to dislodge unnecessary features and send the remaining features to a flexible naive Bayesian sorter for malfunction diagnoses. In Herp et al. [12], a method is proposed to establish a fault-diagnosis model by learning fault samples, assuming that the error features picked up from SCADA (Supervisory Control

And Data Acquisition) data compliance a Gaussian distribution in the characteristic space. Wang D. [13] present a method for improving wavelet filtering by combining infographics and Bayesian inference to confirm the best wavelet argument and apply to malfunction diagnoses. In Li et al. [14], in the process of fault feature extraction, the importance of different signals is optimized by particle swarm optimization. Yu et al. [15] propose an error-feature collection means based on Mean Multigrain Decision Theory Rough Sets (MMGDTRS) and Non-Naive Bayes Classifier (NNBC). Li et al. [16] present a new first-rank Bayesian command method for predicting early failure of gear-shaft systems with locally observable degradation and random failure. A polybasic Bayesian command strategy on account of Hidden Semi-Markov Model (HSMM) is proposed. In Liu et al. [17], a state-monitoring method of rolling bearings based on hybrid generalized HMM is introduced, which uses interval value features to effectively identify and classify the state in the machine process. In Gan and Jiao [18], a malfunction diagnoses means of wavelet transform gearbox on account of ameliorated inheritance arithmetic radio frequency sorter is proposed. Li et al. [19] introduced a malfunction diagnoses means for gearboxes on account of deep radio frequency integration of aural and oscillation signals. Han and Jiang [20] use VMD to acquire eigenvectors and send them to RF for fault diagnosis. Qin [21] welded Ensemble Empirical Mode Decomposition (EEMD) and RF for malfunction diagnoses. Verellen et al. [22], aiming at the detection of bearing faults in rotating machinery, propose a non-invasive acoustic signal-monitoring system based on a sparse microphone array. Traditional vibration analysis uses accelerometers, which are touch sensors that need to be attached to the component under investigation. Smieja et al. [23] proposed an interesting non-contact vibration monitoring method in which image processing is used. Cao et al. [24] proposed a pipeline robot fault diagnosis system based on sound-signal recognition, which transmits the sound signal collected by the storage sensor to the upper computer for fault diagnosis, and the test has achieved good results. Suman et al. [25] proposed an acoustic signal mode-determination algorithm based on adaptive Kalman filtering and MFCC, which can effectively detect vehicle health status by using acoustic signals to detect vehicle mechanical faults. Rakesh Kumar et al. [26] established a rainforest species audio signal-recognition model based on the combination of long short-term memory (LSTM) and convolutional neural network (CNN). The models are combined to achieve a high-accuracy, low-loss detection method. Zhuo et al. [27] proposed a program for on-line diagnosis of steel truss structures using sound signals, and proposed an improved offline database-guided response power and phase transformation method. Experiments show that this method can achieve accurate positioning in strong noise environments, and the amount of computation is smaller.

In this paper, the audio-signal monitoring of power equipment is studied deeply. At present, most sound-signal-processing technologies are based on the receiving frequency range of human ear mechanism. The existing technologies lead to many high- and low-frequency sound signals beyond the range of the human ear not being effectively utilized, resulting in the loss of a large number of effective signal data. However, even if the whole-frequency-band signal-extraction method is adopted, the characteristics of signals are difficult to separate from each other, and the extraction is difficult. The essential reason for these problems is that the coverage of sound signals is extremely wide, so the difficulty of recognition is greatly increased [28]. It can be seen that the traditional sound signal-processing technology has considerable limitations. In order to solve this problem, we took another analytical way of thinking: no longer the traditional method, but the audio-processing problem transferred to the field of image processing. As a result, this paper proposes a power equipment based on wavelet transform voice-fault identification analysis method, in which the access to the audio signal by DWT abstracts the wavelet coefficient of sound. The time-frequency diagram and wavelet coefficient diagram of sound signal are output, and the method of machine learning [29] is applied to analyze sound information from the perspective of image texture. In this method, the whole frequency band of sound signal is extracted without any filtering, and then the sound signal is translated into image processing, which can effectively avoid the loss of information data and make use of the advantages of image recognition for classification.

2. Audio Signal Analysis Based on Wavelet Transform

The overall structure of the research idea is shown in Figure 1. This paper studies the feature extraction method of six kinds of power equipment sounds collected by a 96-channel handheld audio imager. Firstly, we can analyze the audio pre-processing method based on Wavelet and Hamming window, and then we can obtain the audio pre-processing device with different image segmentation coefficients based on Wavelet and Hamming window, and then we can obtain the audio pre-processing device with different image segmentation coefficients; finally, based on this result, we use HOG + SVM method to classify and predict different devices, and find that it has a high recognition rate.

- Preprocessing: the digital strainer is used to preemphasize the audio signal, determine the frame length and frame shift of each sound signal, and the Hamming window is used to filter the sound signal by framing and windowing to obtain multi-frame sound signals;
- Wavelet analysis: by obtaining separate sound signal samples of power equipment through preprocessing operation, we can analyze the characteristics of the sound signal, select an appropriate wavelet function to carry out wavelet transform on the sound signal, and obtain the time wavelet coefficient diagram of each audio signal sample;
- Image processing: considering that the wavelet coefficient image obtained in the above steps contains a large number of image features, this study first uses SSIM (Structural Similarity) image processing method to calculate the similarity between wavelet coefficient images of sound signals of different devices and the same device, so as to verify the feasibility of image classification.
- HOG + SVM: extract the hog feature of the obtained wavelet coefficient image, and substitute the extracted feature into the SVM classifier for multi-classification training, so as to achieve the purpose of classification and prediction of the existing image.

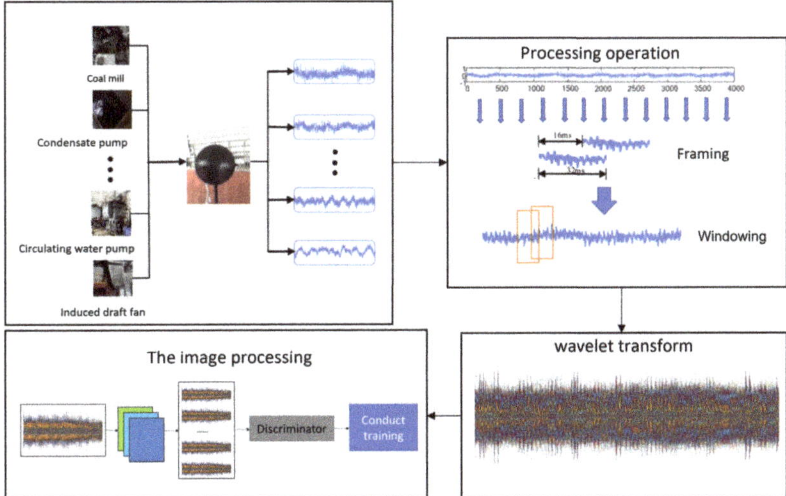

Figure 1. The overall idea of the experimental process. The figure includes the power equipment sound-field-acquisition module, the sound signal preprocessing module, the wavelet transform output image module and the image-processing module.

2.1. Sound Signal Preprocessing

The voice signals collected by the sound imager may have problems such as aliasing, high-order harmonic distortion and high frequency. Before analyzing the sound signals of field equipment, we carry out pre-weighting, framing, windowing and other preprocessing operations so that the signals procured by pursuant voice processing are more consistent and smooth as far as possible, allowing us to afford high-quingity parameters for signal parameter collection and further sound signal processing quality. The specific steps of sound signal preprocessing are as follows:

- Slice. In order to unify the duration of the sound sample, the sound signal of the whole section of audio is segmented into 1 s as a sound sample;
- Pre-emphasis. In order to flatten the spectrum of the sound signal, the spectrum can be calculated with the same structural return loss in the low-frequency to high-frequency band, and the sound signal of each sample is pre-emphasized. Pre-emphasis processing means that the sound signal passes through a high clear strainer:

$$H(z) = 1 - \mu z^{-1} \tag{1}$$

  where in $0.9 < \mu < 1.0$, is taken as 0.97 in this paper.
- Normalization. Normalize the spectrum of the preprocessed sound signal to reduce the difference in the frequency range of different types of sound:

$$X = \frac{X - \min(X)}{\max(X) - \min(X)} \tag{2}$$

- Framing and windowing. The sound signal is stable in a short time. The short-time length is generally 10–30 ms. In order to facilitate feature analysis, the sound signal needs to be processed in frames. For purpose of ensuring the smooth conversion between two adjacent frames, the frame signal needs to be superimposed, and then each frame is multiplied by a window function of a certain length for windowing and filtering. In this paper, Hamming window is adopted, and the window function is shown in Formula (3):

$$0.54 - 0.46\cos\left(\frac{2\pi n}{N-1}\right)(0 \leq n \leq N-1) \tag{3}$$

2.2. Feature Extraction of Audio Signal Based on Wavelet Transform

Wavelet transform is an important time-frequency analysis approach that combines the time-domain characteristics and frequency-domain characteristics of signals.

2.2.1. Definition of Wavelet Function

The application of wavelet analysis in signal and picture compression is a crucial side of the application of wavelet analysis. It has the characteristics of high compression ratio and fast compression speed. After compression, it can not only keep the traits of the signal and image unvaried, but also resist the interference in transmission. The definition formula is as follows:

$$W_f(a, b) = \frac{1}{\sqrt{a}} \sum_{-\infty}^{+\infty} f(x)\phi\left(\frac{x-b}{a}\right)dx \tag{4}$$

Take the function $\phi(x)$ of the basic wavelet as the displacement b, and make the inner product with the signal $f(x)$ to be analyzed under different scales a, with the transformation of a, b the wavelet transform has the traits of multi-resolution.

2.2.2. Wavelet Sequence

$\psi(t) \in L^2(R)$, $\psi(t)$ is called a basic wavelet and mother wavelet, where $L^2(R)$ refers to the mean square integrable space. Wavelet must meet:

$$\sum_{-\infty}^{\infty} \psi(t)dt = 0 \quad (5)$$

This is also the meaning of "wavelet". After scaling and translating the generating function, the wavelet sequence can be obtained:

$$\psi(a,b)(t) = \frac{1}{\sqrt{a}}\psi\left(\frac{t-b}{a}\right) \quad (6)$$

$(a, b \in R, a \neq 0)$ a, b where a, b is the expansion factor and translation factor, respectively.

2.3. *SSIM-Based Image Processing Method*

2.3.1. Definition

Unartificial images have a sehr hoch configuration, especially in the case of spatial similarity, there is a high associations between the pixels of the image. Such associations port crucial information about the configuration of objects in the optical scenario. What we are talking about is finding a more straight method to contradistinguish the configuration of a fuzzy image with that of a reference image.

Structural similarity is a measure of how similar two images are. The SSIM value is between 0 and 1, and the larger its value, the smaller the difference between the images. The definition of SSIM is as in Equation (1) Structural similarity. From the standpoint of image formation, configurational information is defined as a reflection scene that is isolated of brightness and contrast, and the image is modeled by three different factors: brightness, contrast and structure.

Function definition:

$$SSIM(x,y) = [l(x,y)]^{\alpha}[c(x,y)]^{\beta}[s(x,y)]^{\gamma} \quad (7)$$

where $\alpha, \beta, \gamma > 0$.

The measure of similarity can be realized by the SSIM measuring system, which can be constituted of three comparison elements of brightness, contrast and structure. Next, we define three contrast functions:

Brightness contrast function:

$$l(x,y) = \frac{2\mu_x\mu_y + c_1}{\mu_x^2 + \mu_y^2 + c_1} \quad (8)$$

Contrast function:

$$c(x,y) = \frac{2\sigma_x\sigma_y + c_2}{\sigma_x^2 + \sigma_y^2 + c_2} \quad (9)$$

Structural contrast function:

$$s(x,y) = \frac{\sigma_{xy} + c_3}{\sigma_{xy} + c_3} \quad (10)$$

For the above formula, μ_x, μ_y, stand for the whole pixels of the picture; σ_x, σ_y, stand for the criterion differences of picture pixel value; σ_{xy} stand for the convariance of x, y; c_1, c_2, c_3 stand for constants. This is for the purpose of eliminating system fault when the denominator is 0. In practical application, $\alpha = \beta = \gamma = 1$, $c_3 = 0.5c_2$.

2.3.2. Application of SSIM

In image mass evaluation, obtaining the SSIM index of a certain part is better than all. First, the statistical features of images are generally disproportionally distributed *i* then room; second, image deformation varies with the room; third, under standard visual interval, people can centre around one area of the image, therefore the separate processing of a certain part is more in line with the scope of human vision; fourth, the local quality detection can obtain the mapping matrix of image spatial quality changes, and the results can be used for other applications.

Therefore, in the formula above, μ_x, σ_x, σ_{xy} both add an 8×2 square window and traverse the whole image by every pixels. At every procedure of the computation, μ_x, σ_x, σ_{xy} and SSIM values ground on the pixels in the window. Finally, an SSIM index mapping matrix is procured, which is composed of certain part SSIM indexes. However, plain-add window will lead to terrible "blocking" impression of the mapping matrix. To resolve the conundrum, we use the 11×11 meristic Gaussian weighing function $W = \{w_i | i = 1, 2, \cdots, N\}$ as the weighing window, with a par differences of 1.5, and

$$\sum_{i=1}^{N} w_i = 1 \tag{11}$$

Then the approximated value of μ_x, σ_x, σ_{xy} is voiced as:

$$\mu_x = \sum_{i=1}^{N} w_i x_i \tag{12}$$

$$\sigma_x = \left(\sum_{i=1}^{N} w_i (x_i - \mu_x)^2\right)^{\frac{1}{2}} \tag{13}$$

$$\sigma_{xy} = \sum_{i=1}^{N} w_i (x_i - \mu_x)(y_i - \mu_y) \tag{14}$$

Using this windowing means, the mapping matrix can show the capabilities of certain part isotropy, and then use the evenness SSIM index as the evaluation quality of the entire image:

$$MSSIM(x, y) = \frac{1}{MN} \sum_{1}^{M} \sum_{1}^{N} SSIM(x_i, y_i) \tag{15}$$

In the above, x, y are images, x_i, y_i are the locations of certain part SSIM index in the mapping, M, N are the number of local windows.

2.4. HOG Feature Extraction Algorithm

Histogram of Oriented Gradient (HOG) feature is a kind of descriptor that uses computer vision and image processing technology to detect object features. Image features are extracted by calculating and statistical histogram of directional gradient in a specific area of the image. The incorporation of Hog feature extraction and SVM classifier has been diffusely applied in the field of image identification.

Feature Extraction Process

(1) Detection window: Hog cut apart the image through window and block. Mathematically process the pixel values of an area in an image in units of cells. Here, we first introduce the concepts of window, block and cell and the relationship between them.

- Window: divide the image into multiple identical windows according to a certain size and slide;
- Block: divide each window into several same blocks according to a certain size and slide;
- Cell: each window is divided into multiple identical cells according to a certain size, which belong to the feature extraction unit and remain stationary.

(2) Normalized images: Normalization includes gamma and color room normalization. Normalizing the whole image can effectively reduce the influence of lighting conditions. Normalization can also avoid the large proportion of certain part external exposure contribution in picture grain intensity. Standard Gamma compression formula:

$$I(x, y) = I(x, y)^{\gamma} \tag{16}$$

γ takes values based on the effect.

(3) Calculated gradient: Firstly, the gradient value in the horizontal and vertical coordinate orientation is calculated, and the gradient orientation is calculated according to the calculated gradient value. The formula is as follows:

$$G_x(x,y) = H(x+1,y) - H(x-1,y) \tag{17}$$

$$G_y(x,y) = H(x,y+1) - H(x,y-1) \tag{18}$$

For the two formulas $G_x(x,y)$, $G_y(x,y)$, $H(x,y)$ separately stand for the aclinic gradient, perpendicular gradient and pixel value at a specific pixel point of the collected image. The gradient value of amplitude and gradient orientation at pixel (x, y) are:

$$G(x,y) = \sqrt{G_x(x,y)^2 + G_y(x,y)^2} \tag{19}$$

$$\alpha(x,y) = tan^{-1}\left(\frac{G_y(x,y)}{G_x(x,y)}\right) \tag{20}$$

(4) Constructing gradient column diagram: The orientation division is determined by bins (number of divisions). Generally, bins takes 9, and the gradient orientation is cut apart into 9 intervals.

(5) Cell-normalized gradient histogram in the block: the increasing range of gradient intensity is greatly affected by local illumination and foreground–background contrast, so normalization is needed.

(6) Generate hog feature vector: finally, combine all blocks to generate feature vector.

2.5. Support Vector Machines (SVM)

The supervised learning model of support vector machine and its related learning algorithm are widely used in machine learning. It can be used in classification of data and analysis of regression. When given the condition of a set of training specimens, each sample is labeled as one of two different varieties, and the SVM drill algorithm set up a model, deals the new specimens to a certain variety, and constructs an improbability binary linear classifier. The SVM training model represents all specimens as mappings of points in space, and divides the specimens with a wide and obvious gap. The new specimens are then mapped into the same room and their categories predicted.

3. Experimental Result

Firstly, select the working sound of six types of equipment under a fixed working condition collected from the power plant, the sampling frequency is 16,000 Hz, and the fixed 1 s is the cycle for segmentation; The sound sample data set information of six types of equipment is shown in Table 1:

Table 1. sound samples.

Sample Type	Number of Samples	Total Number of Samples
Oil supply pump	200	1200
Connecting shaft	200	
Condensate pump	200	
Coal mill	200	
Induced draft fan	200	
Circulating water pump	200	

After segmentation, the 40 s audio signal of one of the six devices is selected for wavelet transform to obtain the time wavelet coefficient diagram, as shown in the following figures.

From the above image results in Figure 2, it can be seen that there are great differences in time wavelet coefficient images between different devices, and the image features are obvious. Based on this result, we intercept the other four 40 s sound data of each device and output their time wavelet coefficient diagrams. According to the obtained images, we found that the similarity of wavelet coefficient images of a device in different periods is very high, but the feature distinction between different devices is still obvious. Therefore, we took out three images of each device for intra-class and inter-class similarity comparison, and the results are shown below.

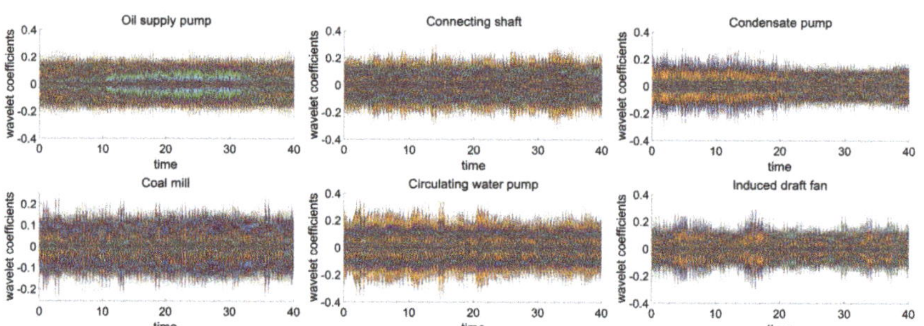

Figure 2. Sample image. The abscissa represents time and the ordinate represents wavelet coefficients.

It can be seen from the Figure 3 that the signal similarity of the same equipment in different periods is generally higher than that between different equipment. Based on the above similarity-matching results, we divide the time wavelet coefficient graphs obtained by each equipment into five different time periods into two groups, one group of four graphs as the training set and the other group of one graph as the test set. In this way, a total of 24 training samples and 6 test samples of 6 types of samples are obtained. The test samples are predicted and classified by using hog feature-extraction algorithm and SVM multi-classification training. The results are shown in Table 2 and Figure 4 below.

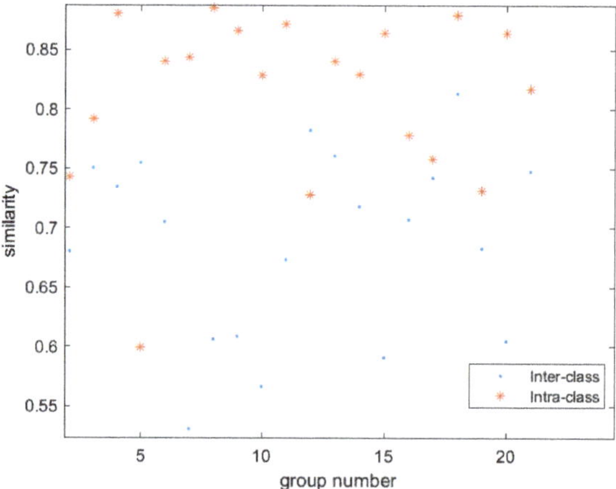

Figure 3. Scatter diagram of image similarity. The abscissa represents the number of sample groups, and the ordinate represents the sample similarity.

Table 2. Classification accuracy of raw data.

Sample Type	Single Class Accuracy	Overall Accuracy
Oil supply pump	80%	90%
Connecting shaft	100%	
Condensate pump	100%	
Coal mill	80%	
Induced draft fan	100%	
Circulating water pump	80%	

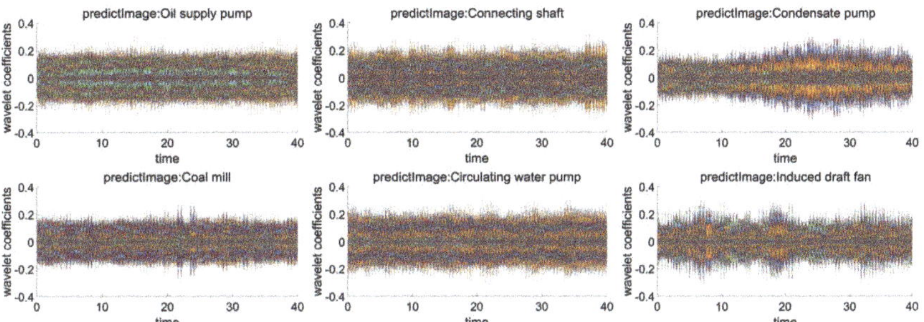

Figure 4. Original sound classification of equipment. The abscissa represents time, and the ordinate represents wavelet coefficients.

In the field of power production, it is difficult to completely eliminate the noise interference in the extraction process of power equipment sound. Therefore, we add Gaussian white noise to the original power equipment sound signal as interference to verify the accuracy and feasibility of this method. Through experiments, we find that when 10 dB Gaussian white noise is added, the characteristics of the time wavelet coefficient diagram of each equipment are not obvious, so it is difficult to distinguish the equipment, When 20 dB Gaussian white noise is added, the characteristics of each equipment in the time wavelet coefficient diagram appear again. Therefore, we process and classify the sound signal added with 20 dB Gaussian white noise. The results are shown in Table 3 and Figure 5 below.

Table 3. Add white noise data classification accuracy.

Sample Type	Single Class Accuracy	Overall Accuracy
Oil supply pump	80%	87%
Connecting shaft	100%	
Condensate pump	100%	
Coal mill	60%	
Induced draft fan	100%	
Circulating water pump	80%	

It can be seen from the experimental results that when white Gaussian noise is affiliated to the sound signal of the equipment, the features of the images of some equipment become more difficult to distinguish, and the recognition accuracy of the image is slightly decreased, but the overall recognition accuracy is high, and the classification effect is obvious. By adding white Gaussian noise of different decibels, it is not difficult to find that noises of different decibels have different degrees of influence on the sound signal of the equipment, which is intuitively reflected in the wavelet coefficient–time diagram, making it more difficult to distinguish image features and equipment identification and classification. Compared with the traditional power equipment sound-recognition method, the advantages of the image processing-based power equipment sound-recognition method proposed in this paper lie in the use of the full frequency range of the sound signal and the more delicate feature expression. For example, a sound-recognition algorithm for substation equipment based on harmonic characteristics and vector quantization was proposed by Dong et al. [30]. The sampled sound signal of power equipment takes the 27th harmonic within 0–1300 Hz as the feature vector, so there will be a lot of noise. The sound data is not used, and the sound features are difficult to express in detail and comprehensively, which will have a certain impact on the accuracy of the results.

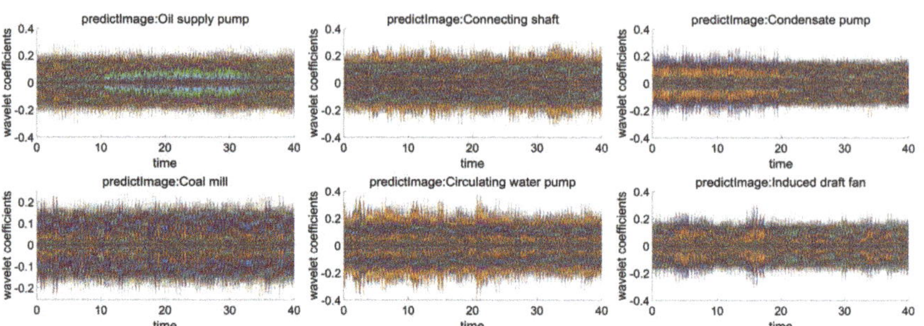

Figure 5. Equipment classification after adding 20 dB Gaussian white noise. The abscissa represents time, and the ordinate represents wavelet coefficients.

4. Conclusions

In this paper, aiming at the sound of six types of thermal power plant power system equipment collected from the scene, the wavelet coefficient–time map of the equipment is obtained through wavelet transformation, and the audio signal is translated into image processing. SSIM algorithm is used to calculate the same at different times and for different equipment, and the image similarity between them can draw a clear difference in terms of image characteristics, which can be used in the classification. Based on this judgment, the obtained images were classified by HOG + SVM fusion method, and 10 dB and 20 dB Gaussian white noise were added to the audio signal, respectively. It was found that noises of different decibels had different degrees of influence on the sound signal of the equipment, and the difficulty of distinguishing the features of the wavelet coefficient–time graph would be improved. Under the influence of 10 dB noise, the characteristic of the wavelet coefficient–time diagram of the equipment is not obvious and difficult to distinguish, but under the influence of 20 dB noise, the difficulty of distinguishing the characteristic of the wavelet coefficient–time diagram of equipment is increased, but the classification effect is good. The experimental results show that the recognition method of sound translation image processing, which is different from the traditional sound-recognition method, has better practical feasibility. The limitation of this paper is that the number of available audio samples is limited, and there is not enough data for training samples. Moreover, only the image obtained by wavelet transform is considered, and whether the image obtained by other methods has better feature distinguishability has not been studied deeply. In the future, we can explore more methods to express characteristic images of sound signals, and continue to study the optimal method of sound signal recognition based on image processing.

Author Contributions: Conceptualization, K.B.; methodology, K.B. and Z.C.; software, Z.C.; validation, Y.Z. (Yong Zhou); formal analysis, Y.Z. (Yong Zhou); resources, W.B.; data curation, W.B.; writing—original draft prepara-tion, Z.C.; writing—review and editing, K.B.; visualization, N.Z.; supervision, Y.Z. (Yongjie Zhai); project adminis-tration, Y.Z. (Yongjie Zhai); funding acquisition, Y.Z. (Yongjie Zhai). All authors have read and agreed to the published version of the manuscript.

Funding: This research received no external funding.

Institutional Review Board Statement: Not applicable.

Informed Consent Statement: Not applicable.

Data Availability Statement: Not applicable.

Conflicts of Interest: The authors declare no conflict of interest.

References

1. Liang, L.; Liu, S.; Li, Y.; Zhong, M.; Li, Y. Distributed fault detection and isolation for power system. *Int. J. Robust Nonlinear Control* **2021**, *32*, 2143–2158. [CrossRef]
2. Bakhtadze, N.; Yadikin, I. Analysis and Prediction of Electric Power System's Stability Based on Virtual State Estimators. *Mathematics* **2021**, *9*, 3194. [CrossRef]
3. Peng, J.; Yang, P.; Liu, Z.; Sun, G. Double-Fed Wind Power System Adaptive Sensing Control and Condition Monitoring. *J. Sens.* **2021**, *2021*, 5753947. [CrossRef]
4. Liu, R.; Yang, B.; Zio, E.; Chen, X. Artificial intelligence for fault diagnosis of rotating machinery: A review. *Mech. Syst. Signal Process.* **2018**, *108*, 33–47. [CrossRef]
5. Yang, Z.; Diao, C.; Li, B. A Robust Hybrid Deep Learning Model for Spatiotemporal Image Fusion. *Remote Sens.* **2021**, *13*, 5005. [CrossRef]
6. Cao, P.; Zhang, S.; Tang, J. Preprocessing-Free Gear Fault Diagnosis Using Small Datasets With Deep Convolutional Neural Network-Based Transfer Learning. *IEEE Access* **2018**, *6*, 26241–26253. [CrossRef]
7. Kafeel, A.; Aziz, S.; Awais, M.; Khan, M.A.; Afaq, K.; Idris, S.A.; Mostafa, S.M. An Expert System for Rotating Machine Fault Detection Using Vibration Signal Analysis. *Sensors* **2021**, *21*, 7587. [CrossRef]
8. Ribeiro, L.C.; Bonaldi, E.L.; de Oliveira, L.E.L.; da Silva, L.E.B.; Salomon, C.P.; Santana, W.C.; Silva, J.G.B.; Lambert-Torres, G. Equipment for Predictive Maintenance in Hydrogenerators. *AASRI Procedia* **2014**, *7*, 75–80. [CrossRef]
9. Song, Z.; Zhang, Z.; Jiang, Y.; Zhu, J. Wind turbine health state monitoring based on a Bayesian data-driven approach. *Renew. Energy* **2018**, *125*, 172–181. [CrossRef]
10. Li, Q.; Hu, W.; Peng, E.; Liang, S.Y. Multichannel Signals Reconstruction Based on Tunable Q-Factor Wavelet Transform-Morphological Component Analysis and Sparse Bayesian Iteration for Rotating Machines. *Entropy* **2018**, *20*, 263. [CrossRef]
11. Yu, J.; Bai, M.; Wang, G.; Shi, X. Fault diagnosis of planetary gearbox with incomplete information using assignment reduction and flexible naive Bayesian classifier. *J. Mech. Sci. Technol.* **2018**, *32*, 37–47. [CrossRef]
12. Herp, J.; Ramezani, M.H.; Bach-Andersen, M.; Pedersen, N.L.; Nadimi, E.S. Bayesian state prediction of wind turbine bearing failure. *Renew. Energy* **2018**, *116*, 164–172. [CrossRef]
13. Wang, D. An extension of the infograms to novel Bayesian inference for bearing fault feature identification. *Mech. Syst. Signal Process.* **2016**, *80*, 19–30. [CrossRef]
14. Li, K.; Zhang, Q.; Wang, K.; Chen, P.; Wang, H. Intelligent Condition Diagnosis Method Based on Adaptive Statistic Test Filter and Diagnostic Bayesian Network. *Sensors* **2016**, *16*, 76. [CrossRef]
15. Yu, J.; Ding, B.; He, Y. Rolling bearing fault diagnosis based on mean multigranulation decision-theoretic rough set and non-naive Bayesian classifier. *J. Mech. Sci. Technol.* **2018**, *32*, 5201–5211. [CrossRef]
16. Li, X.; Makis, V.; Zuo, H.; Cai, J. Optimal Bayesian control policy for gear shaft fault detection using hidden semi-Markov model. *Comput. Ind. Eng.* **2018**, *119*, 21–35. [CrossRef]
17. Liu, J.; Hu, Y.; Wu, B.; Wang, Y.; Xie, F.; Wang, X. A Hybrid Generalized Hidden Markov Model-Based Condition Monitoring Approach for Rolling Bearings. *Sensors* **2017**, *17*, 1143. [CrossRef] [PubMed]
18. Gan, H.; Jiao, B. Fault Diagnosis of Wind Turbine's Gearbox Based on Improved GA Random Forest Classifier. *DEStech Trans. Eng. Technol. Res.* **2018** , 206–210. [CrossRef]
19. Li, C.; Sanchez, R.V.; Zurita, G.; Cerrada, M.; Cabrera, D.; Vásquez, R.E. Gearbox fault diagnosis based on deep random forest fusion of acoustic and vibratory signals. *Mech. Syst. Signal Process.* **2016**, *76*, 283–293. [CrossRef]
20. Han, T.; Jiang, D. Rolling Bearing Fault Diagnostic Method Based on VMD-AR Model and Random Forest Classifier. *Shock Vib.* **2016**, *2016*, 5132046.
21. Qin, X.; Li, Q.; Dong, X.; Lv, S. The Fault Diagnosis of Rolling Bearing Based on Ensemble Empirical Mode Decomposition and Random Forest. *Shock Vib.* **2017**, *2017*, 2623081. [CrossRef]
22. Verellen, T.; Verbelen, F.; Stockman, K.; Steckel, J. Beamforming Applied to Ultrasound Analysis in Detection of Bearing Defects. *Sensors* **2021**, *21*, 6803. [CrossRef] [PubMed]
23. Śmieja, M.; Mamala, J.; Prażnowski, K.; Ciepliński, T.; Szumilas, Ł. Motion Magnification of Vibration Image in Estimation of Technical Object Condition-Review. *Sensors* **2021**, *21*, 6572. [CrossRef]
24. Cao, H.; Yu, J.; Wang, Y.; Zhang, L.; Kim, J. A Fault Diagnosis System for a Pipeline Robot Based on Sound Signal Recognition. *Sensors* **2022**, *22*, 3275. [CrossRef] [PubMed]
25. Suman, A.; Kumar, C.; Suman, P. Early detection of mechanical malfunctions in vehicles using sound signal processing. *Appl. Acoust.* **2022**, *188*, 108578. [CrossRef]
26. Kumar, R.; Gupta, M.; Ahmed, S.; Alhumam, A.; Aggarwal, T. Intelligent Audio Signal Processing for Detecting Rainforest Species Using Deep Learning. *Intell. Autom. Soft Comput.* **2022**, *31*, 693–706. [CrossRef]
27. Zhuo, D.; Cao, H. Damage identification of bolt connection in steel truss structures by using sound signals. *Struct. Health Monit.* **2022**, *21*, 501–517. [CrossRef]

28. Birch, B.; Griffiths, C.A.; Morgan, A. Environmental effects on reliability and accuracy of MFCC based voice recognition for industrial human-robot-interaction. *Proc. Inst. Mech. Eng. Part J. Eng. Manuf.* **2021**, *235*, 1939–1948. [CrossRef]
29. Liu, C.L.; Qi, W.X. Research on Fault Diagnosis Method of Wind Turbine Based on Wavelet Analysis and LS-SVM. *Adv. Mater. Res.* **2013**, *2479*, 724–725. [CrossRef]
30. Li, D.S.; Zhou, Z.Q.; Zhang, C.; Du, P.; Hu, Y.R. Sound Recognition Algorithm for Power Devices Based on Substation Inspection Robots. *Appl. Mech. Mater.* **2014**, *3360*, 1139–1144. [CrossRef]

Article

Life Prediction under Charging Process of Lithium-Ion Batteries Based on AutoML

Chenqiang Luo [1,*], Zhendong Zhang [1,*], Dongdong Qiao [2], Xin Lai [1], Yongying Li [1] and Shunli Wang [3,4]

[1] College of Mechanical Engineering, University of Shanghai for Science and Technology, Shanghai 200093, China; laixin@usst.edu.cn (X.L.); 15901865905@163.com (Y.L.)
[2] School of Automotive Studies, Tongji University, Shanghai 201804, China; qiaodongdong@tongji.edu.cn
[3] College of Electrical Engineering, Sichuan University, Chengdu 610065, China; wangshunli@swust.edu.cn
[4] School of Information Engineering, Southwest University of Science and Technology, Mianyang 621010, China
* Correspondence: 713luochq@163.com (C.L.); usstzzd@usst.edu.cn (Z.Z.)

Abstract: Accurate online capacity estimation and life prediction of lithium-ion batteries (LIBs) are crucial to large-scale commercial use for electric vehicles. The data-driven method lately has drawn great attention in this field due to efficient machine learning, but it remains an ongoing challenge in the feature extraction related to battery lifespan. Some studies focus on the features only in the battery constant current (CC) charging phase, regardless of the joint impact including the constant voltage (CV) charging phase on the battery aging, which can lead to estimation deviation. In this study, we analyze the features of the CC and CV phases using the optimized incremental capacity (IC) curve, showing the strong relevance between the IC curve in the CC phase as well as charging capacity in the CV phase and battery lifespan. Then, the life prediction model based on automated machine learning (AutoML) is established, which can automatically generate a suitable pipeline with less human intervention, overcoming the problem of redundant model information and high computational cost. The proposed method is verified on NASA's LIBs cycle life datasets, with the MAE increased by 52.8% and RMSE increased by 48.3% compared to other methods using the same datasets and training method, accomplishing an obvious enhancement in online life prediction with small-scale datasets.

Keywords: lithium-ion battery; incremental capacity; automated machine learning; life prediction

Citation: Luo, C.; Zhang, Z.; Qiao, D.; Lai, X.; Li, Y.; Wang, S. Life Prediction under Charging Process of Lithium-Ion Batteries Based on AutoML. *Energies* **2022**, *15*, 4594. https://doi.org/10.3390/en15134594

Academic Editor: Quanqing Yu

Received: 26 May 2022
Accepted: 21 June 2022
Published: 23 June 2022

Publisher's Note: MDPI stays neutral with regard to jurisdictional claims in published maps and institutional affiliations.

1. Introduction

With the increasing reduction of fossil energy reserves and severe air pollution, considerable attention has been paid to electric vehicles (EVs) in recent years, which can be energy-saving and environmental-friendly solutions, whereas the traditional automobile industry is a big energy consumer, causing serious exhaust emission [1,2].

Lithium-ion batteries (LIBs) are the ideal energy storage device for EVs, and their safe and feasible application as a power source can contribute to their value in their entire lifespan, which can promote secondary utilization and material recycling, conducting the carbon footprint in the battery production and recycling stages [3]. Hence, the safety and reliability of LIBs in EVs have been spotlighted. However, unlike in the laboratory cycle, the battery performance and the available capacity degrades erratically due to random operation during driving, which could cause underuse or overuse of battery cells, leading to resource waste as well as some potential disasters without accurate remaining useful life (RUL) prediction. Consequently, life prediction and capacity estimation of LIBs are facing challenges, and it is worthwhile devoting much effort to elucidate the battery degradation evolution trend in lifespan, thus, avoiding premature replacement and excessive use.

The typical method of LIBs RUL prediction is usually divided into two categories: model-driven and data-driven methods. The model-driven method exploits the intrinsic aging mechanism and induces complex equations to reflect the reactions process. These models are often established in the theoretical derivation process using mechanism knowledge. Additionally, model parameters are identified through empirical assumptions and mathematical algorithms, such as extended Kalman filter (EKF) [4], expectation maximization (EM) [5], unscented particle filter (UPF) [6], and autoregressive moving average (ARMA) [7]. Nonetheless, it is difficult to predict precisely because of the complex nonlinear process and discrepancy between data distribution and model hypothesis.

The data-driven method has recently received significant attention in LIB's RUL prediction because of easy access to data and the development of machine learning. The raw measured data during operation can serve as the learning model and bridge the implicit gap between the input and output data. More importantly, the successful application of some advanced learning algorithms in machine translation, speech recognition, and computer vision provides a remarkable applying reference for state estimation and RUL prediction in LIBs.

Data-driven methods for RUL prediction are cataloged as machine learning (ML), artificial neural network (ANN), and deep learning (DL). ML: Richardson et al. [8] proposed a regression of the Gaussian process (GP) algorithm for LIBs RUL prediction, with a good performance in long-term forecasting. Yun et al. [9] explored a hybrid prognosis approach for RUL estimation, combining the variational mode decomposition (VMD), autoregressive integrated moving average (ARIMA), and gray model (GM) models for RUL prediction. ANN: Zhang et al. [10] suggested a novel method based on ANN with four layers for state of health (SOH) estimation and RUL prediction using the incremental capacity curve during the constant current discharge phase. Sun et al. [11] proposed a cloud-edge collaborative strategy with state of health (SOH) for capacity estimation and back propagation neural network (BPNN) optimized by a genetic algorithm for capacity prediction. DL: Dong et al. [12] applied the long short-term memory (LSTM) for the RUL prediction, which can solve the gradient exploding problem during iterating. Zraibi et al. [13] pointed out a CNN-LSTM-DNN algorithm for RUL prediction, in which the three hybrid methods respectively play a critical role. Wang et al. [14] proposed a hybrid method combined with a BiLSTM-AM model and a support vector regression (SVR) model for online life prediction, and the collected initial data are updated by SVR, and BiLSTM-AM is used to predict cycle life. Tang et al. [15] decomposed the original data into high- and low-frequency parts precisely through an ensemble empirical mode decomposition, and the parts separately are predicted by DNN and a self-designed LSTM network, named IRes2Net-BiGRU-FC, which showed a high robustness of RUL prediction in both the CC and CV stages.

Currently, most of the research in RUL prediction currently focuses on the application of deep learning and their variant with an intricate network, which can overcome the vanishing and exploding gradient, over-fitting in training, and distortion in long-term dependence through architecture optimization and hyperparameters tuning, which has made outstanding accomplishments. However, some problems also occur, for example, it is difficult to achieve a good training speed and effect for small-scale data in short order due to unmatched model structure and human experts must be deeply involved in every segment of the designing model because of its knowledge- and labor-intensive characteristic. A model with less human intervention and adjustable structure can broaden the exploration of RUL prediction based on the data-driven method.

In recent years, automated machine learning (AutoML) has emerged as a new subarea in machine learning, aiming at tailoring every segment of the machine learning model as a pipeline automatically without requiring human assistance, a combination of automation and ML as defined in Ref. [16]. This model has applied in predicting COVID-19 pandemic [17], Computer Vision [18], Natural Language Process [19], Video Analytics [20], etc. However, existing AutoML research on RUL prediction is just beginning and challenges do emerge, as dealing with the long multivariate time-series problem requires extensive data pre-processing and feasible feature extraction, to ensure that useful information can be accumulated and transmitted. All the three published articles [21–23] proposed RUL prediction of aircraft engines based on AutoML using a simulated turbofan engine degradation open-source dataset from NASA PCoE [24]. Kefalas et al. [21] used a mature architecture TPOT [25] in automatic modeling to develop and optimize machine learning pipelines in an automatic manner, introducing expanding windows to extract statistical features to evaluate the degradation accumulated in the early life of the system. Tornede et al. [22] pointed out a cooperative coevolutionary algorithm, which enlarges the number of pipelines that are explored in a single run, through coevolving the two populations, which are in sub-spaces partitioned by search space into feature extraction and regression methods. Tornede et al. [23] proposed an adaptation of the AutoML tool ML-Plan to automated RUL prediction, integrated an automated feature engineering process transforming time-series data into a standard feature representation, which can deal with a prediction as a standard process. RUL prediction of battery is more challenging since equipment as above runs attaching many sensors to monitor the real-time state, generating more input data of model than battery.

This study aims to develop a life prediction approach based on AutoML using the incremental capacity (IC) curve. The main contributions of this study are summarized as follows:

1. The time-series characteristics in battery constant-current (CC) charging phrase are retained and gathered respectively in curve size by IC analysis as a feature extraction method, which incorporates two healthy indexes (HIs) from inflection point height. Moreover, an IC curve smoothing method based on the Kalman filter (KF) algorithm is also employed to eliminate noise caused by different sampling intervals.
2. The charging capacity of the battery constant-voltage (CV) charging phrase is extracted as another HI directly instead of conversion by the IC method, which ensures the characteristic undistorted transmission in practice. Based on the investigation of the aging mechanism and judgment of correlation analyses, it is proved that the selected features in this study are accurate and typical, which can characterize the aging phenomenon in the entire charging process including the CC and CV phases.
3. The prediction model is established based on AutoML, where an automated pipeline runs in Auto-sklearn architecture with data pre-processing, feature engineering, and automatic modeling. To our knowledge, the model is firstly applied in the RUL prediction of LIBs.
4. The proposed method is verified on an open-source database from NASA PCoE [26], and the results achieve higher accuracy compared with those of other methods. It demonstrates that it can accomplish an obvious performance improvement in online capacity prediction with the small-scale dataset.

The remainder of this paper is prepared as follows: Section 2 introduces the IC analysis and HI extraction. In Section 3, the model based on AutoML is proposed to predict battery RUL. Section 4 presents the experimental results of the proposed method and the valid comparison with other methods. Finally, a conclusion is given in Section 5.

2. IC Analysis and HIs Extraction

2.1. Experimental Dataset

The experimental dataset in this paper is obtained from the NASA Ames Prognostics Center of Excellence, which consists of aging data for 18650 LIBs [26]. The four LIBs B05, B06, B07, and B18 are tested in standard charging and discharging processes under 25 Celsius. The experimental steps of a cycle are as follows and are shown in Figure 1a:

1. Charging process: first, the voltage was raised to 4.2 V under a 1.5 A constant current (CC phase), and then kept charging at 4.2 V constant voltage until the charge current dropped to 20 mA (CV phase).
2. Discharge process: the battery was discharged at a constant current of 2 A until the voltage of B05, B06, B07, and B18 dropped to the cutoff voltage of 2.7 V, 2.5 V, 2.2 V, and 2.5 V, respectively.

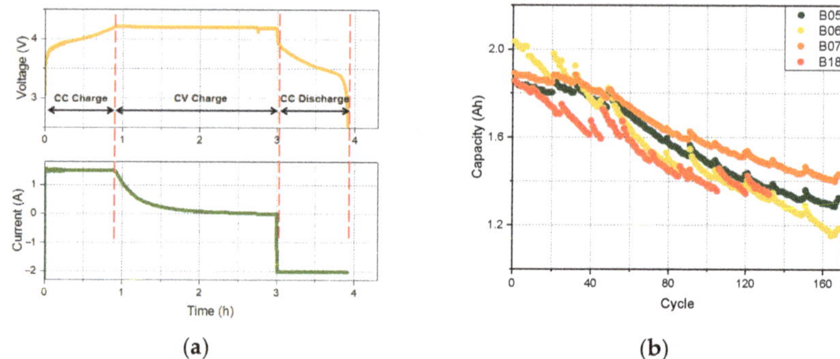

(a) (b)

Figure 1. The experimental steps and capacity degradation profiles: (**a**) The voltage and current in a test cycle; (**b**) Capacity aging trends of the four batteries.

The experiments were halted when the battery capacity decayed to 70%. The tested charge life cycle number of B05, B06, B07, and B18 batteries are 168, 168, 168, and 132 cycles, respectively. The degradation tendencies of battery capacity under different cycle numbers are described in Figure 1b. In this paper, the charging process is selected to study the aging law of LIBs. Figure 2 illustrates the variations of voltage and current of B05 in the CC-CV charging as battery aging. In the CC charging phase, the duration shortens, and the voltage curve moves leftward as battery cycling, which shows the charging capacity in this phase is decreasing. In the CV charging phase, it presents an increasing duration, and the charging capacity showed an opposite trend as in the CC phase.

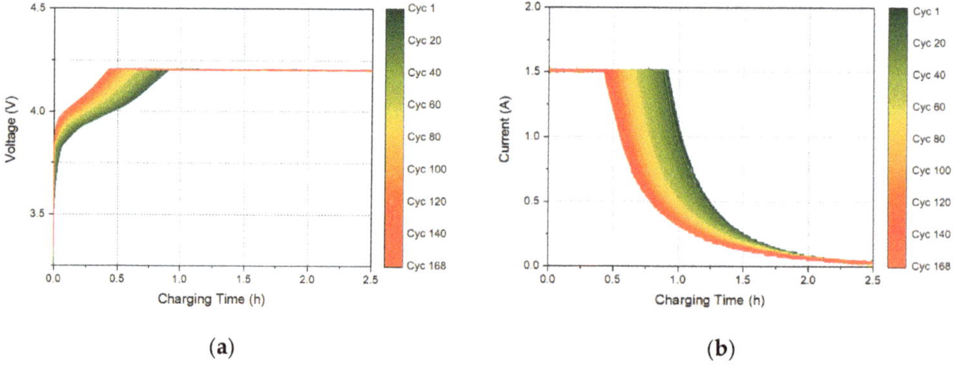

(a) (b)

Figure 2. The variations of voltage and current of B05 in the CC-CV charging: (**a**) Charge voltage; (**b**) Charge current.

2.2. IC Curve and Smoothing Method

The IC curve indicates the change rate of capacity over the voltage evolution during the charging process as an efficient tool to study the variation in the electrochemical properties of LIBs. It has been proved that the batteries with different aging levels have a slight shift in charging voltage or current curve due to the big voltage plateaus during the low-rate cycle [27]. By calculating the derivative of the charging capacity to battery voltage, the IC curve analysis can convert the voltage plateaus into the intuitive and recognizable fluctuation on the IC curve, to detect a gentle change accurately during battery aging [28,29].

The intensity of reactions between electrodes is affected by battery aging during the charging process, where the difference is implicit in the voltage curve but can be reflected in the IC curve as inflection points or even peaks [30]. We can track the battery state and even predict the battery aging trajectory from the inflection points vanishing, decreasing, and shifting, since the slight capacity aging caused by battery degradation can be identified quantitatively by the IC curve [31].

Because the charging capacity is divided by the terminal voltage change within an equal time interval (ETI) and equal voltage interval (EVI) [32], the IC curve can be obtained as shown in Equations (1) and (2).

$$\mathrm{IC}_{\mathrm{ETI}} = \frac{dQ_C}{dV_C} \approx \frac{\Delta Q_C}{\Delta V_C} = \frac{i_C \Delta t}{V_{C,2} - V_{C,1}} \tag{1}$$

$$\mathrm{IC}_{\mathrm{EVI}} = \frac{dQ_C}{dV_C} \approx \frac{\Delta Q_C}{\Delta V_C} = \frac{Q_{C,2} - Q_{C,1}}{\Delta V_C} \tag{2}$$

where Q_C and V_C are the battery charging capacity and battery terminal voltage, respectively. i_C and t are the current and the time in the CC charging phase, which can calculate the charging capacity in the ETI method. $Q_{C,2} - Q_{C,1}$ is the charging capacity in the CV charging phase.

As shown in the green line in Figure 3, the curve calculated by Equation (1) as the sample is polluted by measurement noise owing to impact by the selected interval. If the interval is small, the IC curve will be noisy, and if the interval is large, the IC curve features will become indistinct [32]. It is challenging to catch useful shape features as the peak characteristic is not transparent. In this study, the Kalman filter (KF) is applied as a proper filtering algorithm to improve the curve smoothing.

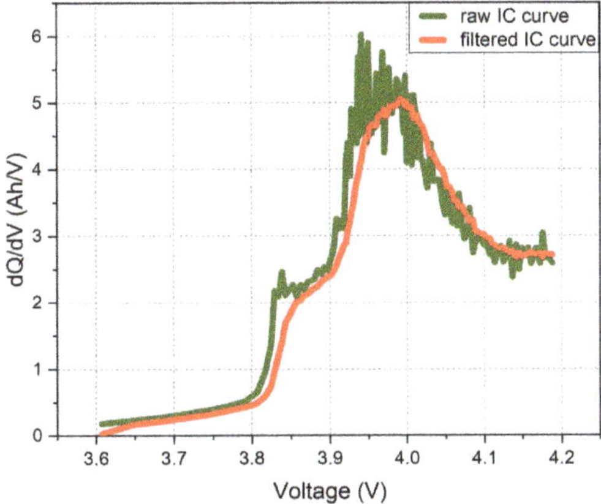

Figure 3. Smoothing results of IC curve.

Firstly, the evolution of $x = \Delta Q/\Delta V$ can be modeled as a random walk with additive Gaussian process noise ω and measurement noise v, then the state equation and observation equation are as follows:

$$\begin{cases} x_k = A_k x_{k-1} + B_k u_k + \omega_k \\ y_k = C_k x_k + D_k u_k + v_k \end{cases} \tag{3}$$

where y_k represents the noise-polluted measurement of x_k, u_k is the external input of the system, ω_k represents the measurement noise, and v_k represents the process noise. Q and R are defined as the covariance of process noise and measurement noise, K_k is the Kalman gain, and P_k is the covariance of estimate value. Then, the filtering algorithm based on the nominal model of Equation (3) can be formulated as:

State and error covariance,

$$\begin{cases} \hat{x}_0^+ = \mathrm{E}(x_0) \\ P_0^+ = \mathrm{E}[(x_0 - \hat{x}_0^+)(x_0 - \hat{x}_0^+)^T] \end{cases} \tag{4}$$

Process and measurement noise,

$$\begin{cases} Q = E(\omega_k \omega_k^T) \\ R = E(v_k v_k^T) \end{cases} \tag{5}$$

State and error covariance time update,

$$\begin{cases} \hat{x}_k^- = A_k \hat{x}_{k-1}^+ + B_k u_k \\ P_k^- = A_k P_{k-1}^+ A_k^T + Q \end{cases} \tag{6}$$

Kalman gain,

$$K_k = P_k^- C_k^T (C_k P_k^- C_k^T + R)^{-1} \tag{7}$$

State and error covariance measurement update,

$$\begin{cases} \hat{x}_k^+ = \hat{x}_{k-1}^- + K_k (y_k - C_k \hat{x}_k^- - D_k u_k) \\ P_k^+ = (I - K_k C_k) P_k^- \end{cases} \tag{8}$$

where u_k is defined as zero due to no external input of the system, and the system state and system output are $x_k = (dQ/dV)_k$ and $y_k = \int i_c dt$. We set $x_0 = 0$ and $P_0 = 1$. The red curve in Figure 3 is the filtered IC curve and the inflection points can be clearly identified. The smoothing method provides a basis for the further development of the HIs extraction.

2.3. *HIs Extraction and Correlation Analysis*

2.3.1. Aging Mechanism Based on IC Analysis

Battery aging is a certainty with corrosion and consumption in the internal material structure of the battery due to electrochemical as well as side reactions in the battery during cycling and storage [33]. According to the research of Ref. [34], the aging mechanisms for LIBs can be categorized into the two main degradation phases: linear degradation phase and accelerated degradation phase. In the linear degradation phase, the battery capacity declines under a linear trend, which is mainly caused by the loss of lithium inventory (LLI), including the formation of SEI film on the surface of the negative electrode, the dilution of electrolyte derived from the side reactions, lithium deposition, and other typical aging mechanisms [35]. In the accelerated degradation phase, the battery capacity is aggravated to decline, where the loss of active material (LAM) emerges as a major factor. The active material is physically damaged and decomposed by the chemical reaction, which affects the intensity of the electrochemical reaction and the transportation of lithium ions between electrodes. Moreover, the products generated by the LLI aging mode, and the polymer decomposed by the electrolyte and lithium deposition can accumulate and be attached to the active material, causing isolation between lithium ions and material as well as material breaking [36,37].

These aging modes are also distinguished in the IC curve. As depicted in Figure 4a, the IC curve in different charging cycles, B05 as the sample, displayed a clear rightward and lower trend along with two obvious inflection points (IP) on the curve, named IP A and IP B. Owe to discrepancy in LIBs internal characteristics, the intensity of inflection is variable. The IC curves of some batteries show slight inflection, and others are inflected into a peak. According to the previous research [32,38] on the degradation mode based on the IC curve feature, it is clear that in terms of LLI and LAM, the intensity of IP A and B will decrease and move toward the higher voltage section during battery cycling, just as Figure 4a. Conversely, IP A and B evolve in opposite trends, and the intensity of IP A is more influenced by LAM mode than LLI mode, and the intensity of IP B mainly depends on LLI mode. Furthermore, for the scenario of EV driving, the battery works in a linear degradation phase as the EOL is most defined as the range from 70% to 80%, so IP B can be more recommended to be the indicator containing aging information than IP A.

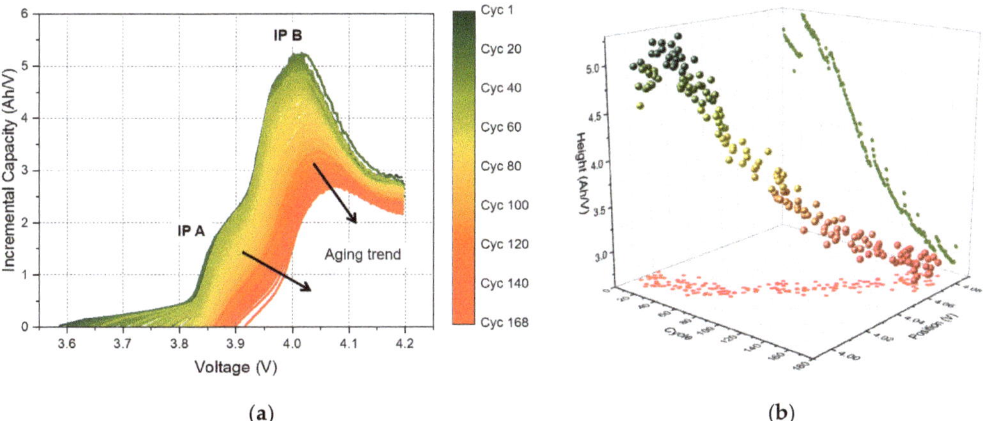

Figure 4. IC curve and HIs of B05 in different charging cycles: (**a**) IC curve; (**b**) Height and position of IP B as two HIs.

2.3.2. Extraction of HIs

In the driving scenario, the battery cannot be discharged under constant current conditions as in the laboratory, which depends on the unpredictable load demand during driving. It is hard to calculate the capacity through the Ampere hour integral method and capture characteristic aging parameters by the discharging curves. Conversely, the charging process is constant due to the regular charging strategy, where the slight shift can be identified. In this study, the IC curve is calculated through the ETI method for the CC charging process as Section 2.2, using charging voltage and CC charging current data. The height of IP B as shape feature characteristic of IC curve and corresponding voltage standing for inflection point position are selected as two HIs for battery RUL prediction, named F1 and F2. The evolution trends of the two HIs with different cycles, B05 still as a sample, are illustrated in Figure 4b.

Compared to laboratory conditions, the charging process is usually incomplete with commercial chargers due to the driver's habit, but the charging curve still retains the shape characteristics of the CC-CV phase, especially the complete curve shape in the CV charging phase, although the different depth of discharge (DOD) in the different cycle may influence battery charging. Owe to the above-mentioned reasons, the HIs of the CV phase can be extracted directly instead of conversion by the IC method, which ensures the characteristic undistorted transmission in practice.

Similar to the CC phase, there are also some regular shifts like indicators for battery aging in the CV phase as Figure 2 depicted. According to voltage balance Equation (5), because of the increase of polarization voltage U_P and ohmic internal resistance during degradation, U_T reaches a cut-off voltage earlier, and the charging process switches to the CV phase in a shorter time, which can lead to different charging time and charging capacity in every cycle as shown in Figure 5.

$$U_T = U_{OCV} + U_P + U_R \tag{9}$$

where U_T is terminal voltage, U_{OCV} is open-circuit voltage, U_P is polarization voltage, and U_R is ohmic internal resistance voltage.

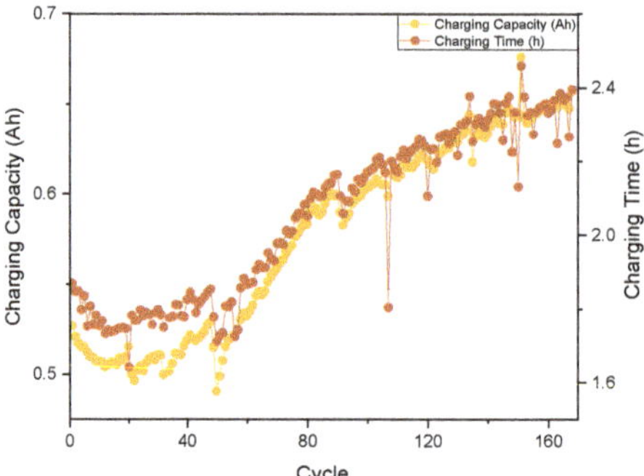

Figure 5. The charging time and capacity profiles of B05 in the CV phase in every cycle.

Hence, the charging capacity of the CV charging phase is chosen as another HI F3 to characterize the capacity degradation, and it can be formulated by the Ampere hour integral method as:

$$Q_{cv} = \int_{t1}^{t2} I_{cv} dt \tag{10}$$

where I_{CV} is the current in the CV charging phase, t_1 and t_2 are the start-stop time of the CV charging phase.

In conclusion, the height of IP B in the IC curve and corresponding voltage standing for the position of inflection point as F1 and F2 in the CC charging phase and the charging capacity of the CV charging phase as F3 are selected as HIs for battery RUL prediction. F1 and F2 can highlight the slight shifts in the voltage plateau phase in the charging process, and F3 represents the charging condition and polarization of the battery. All the HIs reflect the characteristics of the entire battery charging process.

2.3.3. Correlation Analysis of HIs

To further explore the relationship among the three HIs and probe whether all of them can express the change in the battery capacity, the interaction between the HIs and capacity is analyzed by the Pearson correlation and the Spearman correlation. The analysis results are displayed in Table 1 and Figure 6. As the correlation coefficient of F1 (Height of IP B), F3 (Charging capacity of CV phase) is close to 1, and the absolute value of F2 (Position of IP B) also exceeds 0.94, the correlation of model input parameters is quite significant.

Table 1. Results of correlation analysis of three HIs.

HIs	Pearson Correlation	Spearman Correlation
F1	0.9953	0.9889
F2	−0.9506	−0.9455
F3	−0.9895	−0.9737

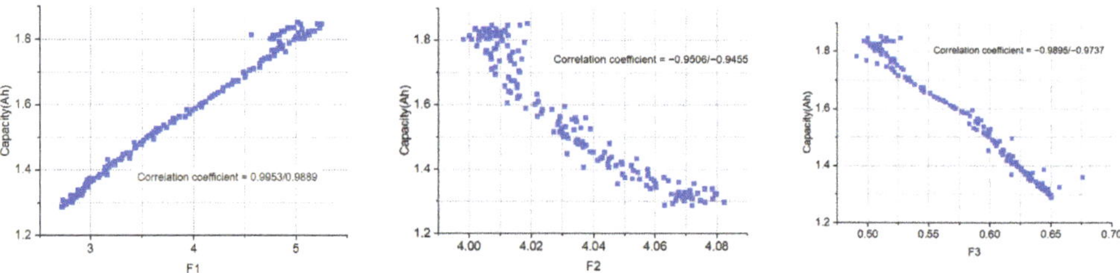

Figure 6. Relationship of HIs and the reference capacity of B05.

3. Online Estimation Based on AutoML

3.1. Description of AutoML Model

We established a novel model based on AutoML, automatically customizing the forecasting pipeline, which consists of data pre-processing, feature engineering, and automatic modeling with less human intervention and trial error manually, covering the complete actions from processing the input data to the deployment of the model.

Each step of data pre-processing, including data cleaning, data augmentation, and data coding, can search the configuration space automatically by some optimization algorithm, such as reinforcement learning and grid search. The pre-processing contributes to input data without polluted noise, avoiding over-fitting of model training and enhancing model robustness. Data pre-processing also involves normalizing, through which the available data can eliminate the effect caused by the different ranges of value in the learning phase. We use the Min-max normalization to map the range of feature S into $[a, b]$ as follows:

$$x' = a + \frac{(x - min(S))(b - a)}{max(S) - min(S)} \tag{11}$$

where x and x' are the value and the transformed value of the feature S.

Feature engineering is to automatically construct features from the data so that subsequent learning models can have good performance, with the segment of feature extraction, feature selection, and feature enhancement. It ensures that the features can exclude the redundant variable and be extracted in an appropriate dimensionality for the feature space.

AutoML aims to solve the so-called CASH problem, the short for combined algorithm selection and hyperparameters optimization [39]. This is essentially the task of selecting or combining the appropriate model for the dataset at hand automatically, along with the various hyperparameters tuned properly in every segment of the pipeline.

The model performance mostly depends on a set of hyperparameters that make up the algorithm. Hyperparameters are tuned specifically to that dataset, with some techniques like Regression Trees, and Gaussian Processes [40]. Bayesian optimization has been applied as a successful candidate for hyperparameter tuning, which fits a probabilistic model to capture the relationship between hyperparameters' setting and their measured performance. Then, the most promising hyperparameter setting is selected and evaluated, as well as updated in the model with the result, finishing an entire iteration [41].

The meta-learning approach is complementary to Bayesian optimization for optimizing a model architecture, which is employed to obtain promising configurations to warmstart Bayesian Optimization. Each model trained on data contributes to the configuration space of hyperparameters cross datasets, even if a model performed poorly. The area of meta-learning [42] follows this common strategy that human experts screen known models by reasoning about the performance of learning algorithms and searching with configuration space. Therefore, meta-learning is applied to select instantiations of the given model frameworks, which tend to perform well on a new dataset, from the knowledge of previous tasks [43]. To characterize explicitly discrepancy in dataset repositories, meta-features are introduced as the searching targets and data depiction, denoting some dimensions, such as Statistical meta-features, PCA meta-features, information-theoretic meta-feature, etc. [43]. They comprise the attributes of each dataset and the parameters of the computing model, such as neural network training weights. More precisely, the meta-learning approach works as follows:

Step 1 For each machine learning dataset in a dataset repository, we evaluated a set of meta-features and used Bayesian optimization to determine and store an instantiation of the given ML framework with strong empirical performance for that dataset.

Step 2 Then, for the given new dataset, we compute its meta-features, sort out all datasets by distance metric among them in meta-feature space and select the stored ML architecture instantiations for the limited nearest datasets for evaluation before starting Bayesian optimization with their results.

In this study, we used Auto-sklearn architecture combined with a meta-learning approach and Bayesian optimization, which is a robust new AutoML system based on scikit-learn, comprising 15 classifiers, 14 feature preprocessing methods, and 4 data preprocessing methods, giving rise to a structured hypothesis space with 110 hyperparameters. The architecture improves the existing AutoML methods by automatically taking into account past performance on similar datasets, and by constructing ensembles from the models evaluated during the optimization [44]. EarlyStopping is used as callbacks to prevent overfitting, which can stop training when the loss did not decrease anymore in each epoch. The network training weights in the current epoch are the final training results.

The flow chart for pipelined-based AutoML is show in Figure 7. Configuration space Λ is built up based on the algorithm repository A in Auto-sklearn architecture, which comprises the hyperparameters controlling each algorithm. Meta-learning searches the existing dataset D_i similar to the new dataset D_{new} in the dataset repository, in which similarity is defined by a distance between two datasets based on meta-features, and initializes a search with the meta-feature F_{Di}. For each dataset, meta-features are only computed on the training set. In contrast to human domain experts, Bayesian Optimization does not use knowledge from previous runs on different datasets and uses the matched F_{Di} to obtain promising configurations λ, in which the model is evaluated with the new meta-feature F_{Dnew} based. The Bayesian Optimization with meta-learning finishes the pipeline including data pre-processing, feature engineering, and classifier selection.

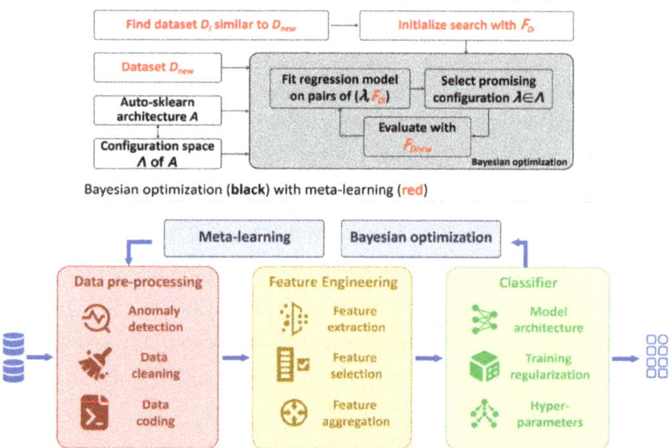

Figure 7. Flow chart for pipeline-based AutoML combined with meta-learning and Bayesian optimization.

3.2. Framework of RUL Prediction Method

The integrated framework for the life prediction approach is described in Figure 8 and divided into offline training and online prediction. In the offline stage, first, the IC curve is calculated in the CC charging phase, in which the two HIs that inflection point height and position are extracted, with the curve smoothing method derived from the KF algorithm. Combined with the charging capacity in the CV phase of every cycle, all three HIs effectively characterize the battery aging of the entire charging process. Then, a novel AutoML model is established with Auto-sklearn architecture, to realize the automatic design of the pipeline. The model hyperparameters are tuned and the search is optimized in the training stage. In the online stage, the extracted three features are directly applied to predict the battery RUL based on the trained AutoML model.

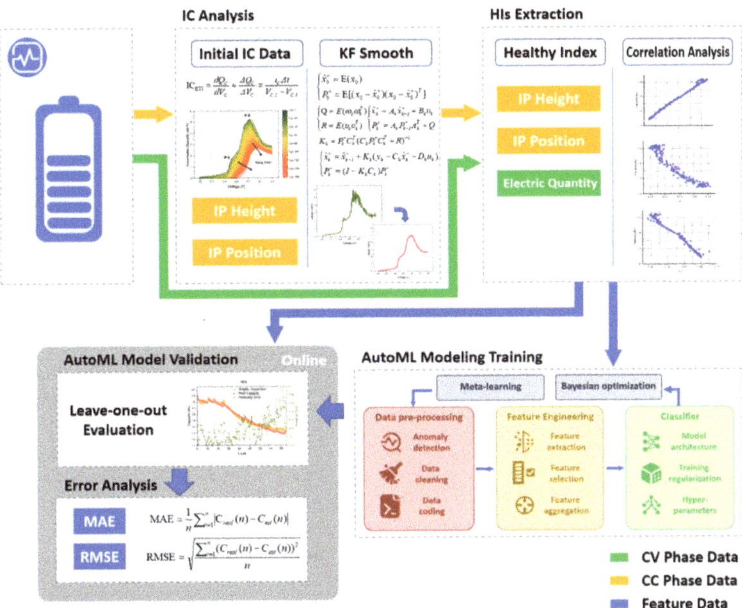

Figure 8. Framework of RUL prediction based AutoML.

4. Results and Discussion

4.1. Evaluation Criteria

In this study, root mean square error (RMSE) and mean absolute error (MAE) indexes are applied to evaluate the performance of the prediction method. The formulae for calculating are as follows:

$$\text{RMSE} = \sqrt{\frac{\sum_{i=1}^{n}(C_{real}(n) - C_{prd}(n))^2}{n}} \tag{12}$$

$$\text{MAE} = \frac{1}{n}\sum_{i=1}^{n}\left|C_{real}(n) - C_{prd}(n)\right| \tag{13}$$

where n is the number of cycles, C_{real} is the real capacity, and C_{prd} is the predicted capacity.

In most experiment settings in data-driven methods, the training set and validation set are usually divided on the same single battery, which can implement the online prediction. In the driving scenario, unlike in the bench test, the BMS cannot predict online capacity without the first 60% to 80% battery running data, so the goal of online prediction for the entire battery aging cycle is hard to achieve. To tackle this, we use the leave-one-out evaluation as Chen et al. [45] applied to evaluate their model: one battery is sampled for validation, and the other three batteries are used for training. A total of four trials are conducted and hyperparameters of the model as well as the average evaluating index over all batteries are determined.

4.2. Prediction Results and Analysis

As the framework showed in Figure 8, the HIs F1, F2, and F3 calculated by raw voltage and current data are used as input of the AutoML model, and the training set and validation set are divided by leave-one-out evaluation.

By means of searching and evaluating, the three methods, poly, rbf, and sigmoid, are used for feature preprocessing, and the five classifiers are selected in the AutoML model, and the classifiers' type and ensembled weight are listed in Table 2. In Table 3 we show the hyperparameters used in this study.

Table 2. The classifiers type and ensembled weight in the AutoML model.

Rank	Classifier Type	Ensembled Weight
1	Gaussian_process	0.50
2	K_nearest_neighbors	0.32
3	Gradient_boosting	0.08
4	Ard_regression	0.08
5	Liblinear_svr	0.02

Table 3. Hyperparameters used in verification.

Hyperparameters	Data Type	Value Range
Initial configurations via meta-learning	int	25
Ensemble size	int	50
Max reserved models	int	50
batch size of training data	int	64
Number of training epochs	int	200
Resampling strategy	cat	Holdout
Model training optimizer	cat	SGD

The capacity prediction and relative error of B05, B06, B07, and B18 are depicted in Figure 9. The red predicted capacity curve approximates the yellow real capacity curve with most prediction errors controlled within 7% in the validation of four batteries, indicating high accuracy and robustness.

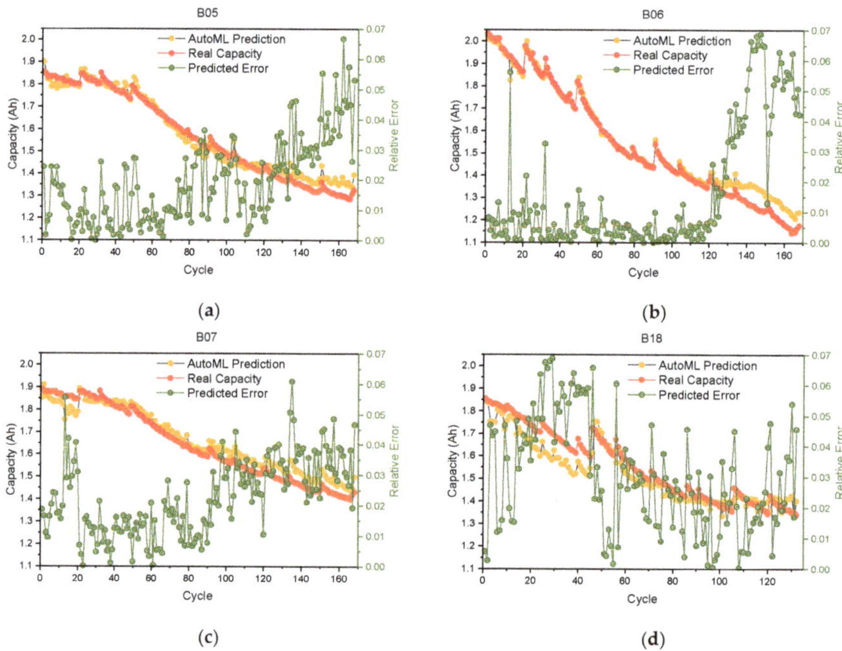

Figure 9. Capacity prediction results and errors of (**a**) B05, (**b**) B06, (**c**) B07, and (**d**) B18.

Compared to the RMSE value in Table 4, the lowest prediction accuracy among the four batteries is B18. Although it is not as good as other batteries, the MAE of B18 is 0.0479, indicating the test accuracy rate reached more than 95%. According to the development requirements of batteries in EVs, the fault threshold is set to 80% of initial capacity, and the life value can also be predicted. The predicted error is 1 and 8 cycles for B06 and B18 respectively. All of the above illustrates that the proposed online prediction method has high accuracy and reliability.

Table 4. Prediction results of four batteries.

	B05	B06	B07	B18
MAE	0.0283	0.0221	0.0361	0.0479
RMSE	0.0337	0.0340	0.0407	0.0573

We conducted the same training and validation for NASA's data and average evaluation index as in Ref. [45], so we can perform a valid benchmark comparison. Table 5 summarizes the RUL prediction results from various methods, and it is obvious that the proposed method achieves a better performance than other methods, which is presented in Ref. [45]. It is obvious that the prediction based on AutoML is more accurate using the same data set and training method, with the MAE increased by 52.8% and RMSE increased by 48.3% than DeTransformer.

Table 5. Comparison of prediction results of AutoML with other ML methods.

	MLP	RNN	LSTM	GRU	Daul-LSTM	DeTransformer	AutoML
MAE	0.1379	0.0749	0.0829	0.0806	0.0815	0.0713	0.0336
RMSE	0.1541	0.0848	0.0905	0.0921	0.0879	0.0802	0.0414

5. Conclusions

In this study, according to our knowledge, we are the first to propose the AutoML model applied in the RUL prediction of LIBs, with the HIs extracted by IC analysis. A smoothing IC curve based on the KF algorithm is employed for HIs extraction and three HIs have been verified to characterize the aging phenomenon in the entire charging process including the CC and CV phases. We proposed a prediction method based on AutoML running in the Auto-sklearn architecture, which can customize the pipeline for specific datasets automatically, overcoming the problem of redundant model information and high computational cost. Then the experiment on NASA's LIBs cycle life dataset verifies the accuracy and robustness. As a next step, we plan to keep on further studies on neural networks in the AutoML model, using neural architecture search to improve pipeline, as well as investigating effective dimensionality optimization techniques for the HIs extraction by IC analysis.

Author Contributions: Conceptualization, C.L. and Z.Z.; methodology, C.L. and D.Q.; software, C.L. and Y.L.; validation, C.L. and Y.L.; formal analysis, C.L. and D.Q.; investigation, C.L.; resources, C.L.; data curation, C.L. and D.Q.; writing—original draft preparation, C.L.; writing—review and editing, C.L.; visualization, C.L.; supervision, Z.Z. and X.L.; project administration, C.L. and S.W.; funding acquisition, Z.Z. and X.L. All authors have read and agreed to the published version of the manuscript.

Funding: This research was funded by the National Natural Science Foundation of China, grant number 51977131.

Institutional Review Board Statement: Not applicable.

Informed Consent Statement: Not applicable.

Data Availability Statement: Not applicable.

Conflicts of Interest: The authors declare no conflict of interest.

References

1. Liu, K. Data-driven health estimation and lifetime prediction of lithium-ion batteries: A review. *Renew. Sustain. Energy Rev.* **2019**, *113*, 109254. [CrossRef]
2. Tian, H.; Qin, P.; Li, K.; Zhao, Z. A review of the state of health for lithium-ion batteries: Research status and suggestions. *J. Clean. Prod.* **2020**, *261*, 120813. [CrossRef]
3. Lai, X.; Chen, Q.; Tang, X.; Zhou, Y.; Gao, F.; Guo, Y.; Bhagat, R.; Zheng, Y. Critical review of life cycle assessment of lithium-ion batteries for electric vehicles: A lifespan perspective. *eTransportation* **2022**, *12*, 100169. [CrossRef]
4. Liao, Z.; Gai, N.; Stansby, P.K.; Li, G. Linear non-causal optimal control of an attenuator type wave energy converter M4. *IEEE Trans. Sustain. Energy* **2020**, *11*, 1278–1286. [CrossRef]
5. Xu, X.; Chen, N. A state-space-based prognostics model for lithium-ion battery degradation. *Reliab. Eng. Syst. Saf.* **2017**, *159*, 47–57. [CrossRef]
6. Miao, Q.; Xie, L.; Cui, H.; Liang, W.; Pecht, M.G. Remaining useful life prediction of lithium-ion battery with unscented particle filter technique. *Microelectron. Reliab.* **2013**, *53*, 805–810. [CrossRef]
7. Jin, X.; Lian, X.; Su, T.; Shi, Y.; Miao, B. Closed-loop estimation for randomly sampled measurements in target tracking system. *Math. Probl. Eng.* **2014**, *2014*, 315908.
8. Richardson, R.R.; Osborne, M.A.; Howey, D.A. Gaussian process regression for forecasting battery state of health. *J. Power Sources* **2017**, *357*, 209–219. [CrossRef]
9. Yun, Z.; Qin, W.; Shi, W.; Ping, P. State-of-health prediction for lithium-ion batteries based on a novel hybrid approach. *Energies* **2020**, *13*, 4858. [CrossRef]
10. Zhang, S.; Zhai, B.; Guo, X.; Wang, K.; Peng, N.; Zhang, X. Synchronous estimation of state of health and remaining useful lifetime for lithium-ion battery using the incremental capacity and artificial neural networks. *J. Energy Storage* **2019**, *26*, 100951.1–100951.12. [CrossRef]
11. Sun, T.; Wang, S.; Jiang, S.; Xu, B.; Han, X.; Lai, X.K.; Zheng, Y. A cloud-edge collaborative strategy for capacity prognostic of lithium-ion batteries based on dynamic weight allocation and machine learning. *Energy* **2022**, *239*, 122185. [CrossRef]
12. Dong, D.; Li, X.Y.; Sun, F.Q. Life prediction of jet engines based on LSTM-recurrent neural networks. In Proceedings of the 2017 Prognostics and System Health Management Conference (PHM-Harbin), Harbin, China, 9–12 July 2017.
13. Zraibi, B.; Okar, C.; Chaoui, H.; Mansouri, M. Remaining useful life assessment for lithium-ion batteries using CNN-LSTM-DNN hybrid method. *IEEE Trans. Veh. Technol.* **2021**, *70*, 4252–4261. [CrossRef]
14. Wang, F.; Zemenu, E.A.; Tseng, C.; Chou, J. A hybrid method for online cycle life prediction of lithium-ion batteries. *Int. J. Energy Res.* **2022**, *46*, 9080–9096. [CrossRef]
15. Tang, T.; Yuan, H. A hybrid approach based on decomposition algorithm and neural network for remaining useful life prediction of lithium-ion battery. *Reliab. Eng. Syst. Saf.* **2022**, *217*, 108082. [CrossRef]
16. Yao, Q.; Wang, M.; Escalante, H.J.; Guyon, I.; Hu, Y.; Li, Y.; Tu, W.; Yang, Q.; Yu, Y. Taking human out of learning applications: A survey on automated machine learning. *arXiv* **2018**, arXiv:1810.13306.
17. Gomathi, S.; Kohli, R.; Soni, M.; Dhiman, G.; Nair, R. Pattern analysis: Predicting COVID-19 pandemic in India using AutoML. *World J. Eng.* **2022**, *19*, 21–28. [CrossRef]
18. Zeng, Y.; Zhang, J. A machine learning model for detecting invasive ductal carcinoma with google cloud AutoML vision. *Comput. Biol. Med.* **2020**, *122*, 103861. [CrossRef]
19. Drori, I.; Liu, L.; Nian, Y.; Koorathota, S.C.; Li, J.; Moretti, A.K.; Freire, J.; Udell, M. AutoML using metadata language embeddings. *arXiv* **2019**, arXiv:1910.03698.
20. Galanopoulos, A.; Ayala-Romero, J.A.; Leith, D.J.; Iosifidis, G. AutoML for video analytics with edge computing. In Proceedings of the IEEE INFOCOM 2021-IEEE Conference on Computer Communications, Vancouver, BC, Canada, 10–13 May 2021.
21. Kefalas, M.; Baratchi, M.; Apostolidis, A.; Herik, D.V.; Bäck, T. Automated machine learning for remaining useful life estimation of aircraft engines. In Proceedings of the 2021 IEEE International Conference on Prognostics and Health Management (ICPHM), Detroit, MI, USA, 7–9 June 2021.
22. Tornede, T.; Tornede, A.; Wever, M.; Hüllermeier, E. Coevolution of remaining useful lifetime estimation pipelines for automated predictive maintenance. In Proceedings of the Genetic and Evolutionary Computation Conference, Lille, France, 10–14 July 2021.
23. Tornede, T.; Tornede, A.; Wever, M.; Mohr, F.; Hüllermeier, E. AutoML for predictive maintenance: One tool to RUL them all. In Proceedings of the IoT Streams for Data-Driven Predictive Maintenance and IoT, Edge, and Mobile for Embedded Machine Learning, Ghent, Belgium, 14–18 September 2020.
24. Chao, M.A.; Kulkarni, C.S.; Goebel, K.F.; Fink, O. Aircraft engine run-to-failure dataset under real flight conditions for prognostics and diagnostics. *Data* **2021**, *6*, 5. [CrossRef]
25. Le, T.T.; Fu, W.; Moore, J. Scaling tree-based automated machine learning to biomedical big data with a feature set selector. *Bioinformatics* **2020**, *36*, 250–256. [CrossRef]

26. Goebel, K.; Saha, B.; Saxena, A.; Celaya, J.R.; Christophersen, J. Prognostics in battery health management. *IEEE Instrum. Meas. Mag.* **2008**, *11*, 33–40. [CrossRef]
27. Qiao, D.; Wei, X.; Fan, W.; Jiang, B.; Lai, X.; Zheng, Y.; Tang, X.; Dai, H. Toward safe carbon–neutral transportation: Battery internal short circuit diagnosis based on cloud data for electric vehicles. *Appl. Energy* **2022**, *317*, 119168. [CrossRef]
28. Feng, X.; Li, J.; Ouyang, M.; Lu, L.; Li, J.; He, X. Using probability density function to evaluate the state of health of lithium-ion batteries. *J. Power Sources* **2013**, *232*, 209–218. [CrossRef]
29. Weng, C.; Jing, S.; Peng, H. A unified open-circuit-voltage model of lithium-ion batteries for state-of-charge estimation and state-of-health monitoring. *J. Power Sources* **2014**, *258*, 228–237. [CrossRef]
30. Xue, N.; Sun, B.; Bai, K.; Han, Z.; Li, N. Different state of charge range cycle degradation mechanism of composite material lithium-ion batteries based on incremental capacity analysis. *Trans. China Electrotech. Soc.* **2017**, *32*, 145–152.
31. Han, X.B. Study on Li-Ion Battery Mechanism Model and State Estimation for Electric Vehicles. Ph.D. Dissertation, Tsinghua University, Beijing, China, 2014.
32. Qiao, D.; Wang, X.; Lai, X.; Zheng, Y.; Wei, X.; Dai, H. Online quantitative diagnosis of internal short circuit for lithium-ion batteries using incremental capacity method. *Energy* **2022**, *243*, 123082. [CrossRef]
33. Vetter, J.; Novák, P.; Wagner, M.R.; Veit, C.; Hammouche, A. Ageing mechanisms in lithium-ion batteries. *J. Power Sources* **2005**, *147*, 269–281. [CrossRef]
34. Dubarry, M.; Berecibar, M.; Devie, A.; Anseán, D.; Omar, N.; Villarreal, I. State of health battery estimator enabling degradation diagnosis: Model and algorithm description. *J. Power Sources* **2017**, *360*, 59–69. [CrossRef]
35. Bloom, I.D.; Cole, B.W.; Sohn, J.; Jones, S.A.; Polzin, E.G.; Battaglia, V.S.; Henriksen, G.L.; Motloch, C.G.; Richardson, R.A.; Unkelhaeuser, T.; et al. An accelerated calendar and cycle life study of Li-ion cells. *J. Power Sources* **2001**, *101*, 238–247. [CrossRef]
36. Dubarry, M.; Truchot, C.; Liaw, B.Y. Synthesize battery degradation modes via a diagnostic and prognostic model. *J. Power Sources* **2012**, *219*, 204–216. [CrossRef]
37. Anseán, D.; Dubarry, M.; Devie, A.; Liaw, B.Y.; Garcia, V.; Viera, J.C.; González, M. Fast charging technique for high power $LiFePO_4$ batteries: A mechanistic analysis of aging. *J. Power Sources* **2016**, *321*, 201–209. [CrossRef]
38. Han, X.; Ouyang, M.; Lu, L.; Li, J.; Zheng, Y.; Li, Z. A comparative study of commercial lithium ion battery cycle life in electrical vehicle: Aging mechanism identification. *J. Power Sources* **2014**, *251*, 38–54. [CrossRef]
39. Thornton, C.; Hutter, F.; Hoos, H.H.; Leyton-Brown, K. Auto-weka: Combined selection and hyperparameter optimization of classification algorithms. In Proceedings of the 19th ACM SIGKDD International Conference on Knowledge Discovery and Data Mining (KDD '13), Chicago, IL, USA, 11–14 August 2013.
40. Nagarajah, T.; Poravi, G. A review on automated machine learning (AutoML) Systems. In Proceedings of the 2019 IEEE 5th International Conference for Convergence in Technology (I2CT), Bombay, India, 29–31 March 2019.
41. Brochu, E.; Cora, V.M.; Freitas, N.D. A tutorial on Bayesian optimization of expensive cost functions, with application to active user modeling and hierarchical reinforcement learning. *arXiv* **2010**, arXiv:1012.2599.
42. Hodgson, J. Metalearning: Applications to data mining. *Comput. Rev.* **2010**, *51*, 217–218.
43. Feurer, M.; Springenberg, J.T.; Hutter, F. Initializing Bayesian hyperparameter optimization via meta-learning. In Proceedings of the Twenty-Ninth AAAI Conference on Artificial Intelligence (AAAI-15), Austin, TX, USA, 25–30 January 2015.
44. Feurer, M.; Klein, A.; Eggensperger, K.; Springenberg, J.T.; Hutter, F. *Automated Machine Learning*, 1st ed.; Springer: Cham, Switzerland, 2019; pp. 113–134.
45. Chen, D.; Hong, W.; Zhou, X. Transformer network for remaining useful life prediction of lithium-ion batteries. *IEEE Access* **2022**, *10*, 19621–19628. [CrossRef]

Article

GIS Partial Discharge Pattern Recognition Based on Time-Frequency Features and Improved Convolutional Neural Network

Jianfeng Zheng [1,2], Zhichao Chen [1], Qun Wang [1], Hao Qiang [1,2] and Weiyue Xu [1,2,*]

[1] School of Mechanical Engineering and Rail Transit, Changzhou University, Changzhou 213164, China
[2] Jiangsu Province Engineering Research Center of High-Level Energy and Power Equipment, Changzhou University, Changzhou 213164, China
* Correspondence: wyxu@cczu.edu.cn

Abstract: Different types of partial discharge (PD) in gas-insulated switchgear (GIS) cause different damage to GIS insulation, correctly identifying the PD type is very important for evaluating the insulation status of GIS. This paper proposes a PD pattern recognition method based on an improved feature fusion convolutional neural network (IFCNN) to fully use the time-frequency features of PD pulses to realize PD pattern recognition. Firstly, the one-dimensional time-domain feature sequence of the PD pulse and the corresponding wavelet time-frequency diagram are applied as inputs. Secondly, the convolutional neural network (CNN) with two parallel channels is used for feature extraction, the extracted fault information is fused, and the shallow features of the wavelet time-frequency diagram are fused to prevent feature loss caused by pooling operation. Finally, the extracted features are sent to the classifier to recognize different types of PD. The discharge data of different types of PD are obtained for testing by experiments and simulation. Compared with 1-D CNN and 2-D CNN under the same specification, the proposed method can mine more potential local features of discharge pulses by fusing the time-frequency features of PD pulses in different dimensions, and improves the recognition accuracy to 95.8%.

Keywords: partial discharge; time-frequency features; wavelet transform; convolutional neural network; pattern recognition

Citation: Zheng, J.; Chen, Z.; Wang, Q.; Qian, H.; Xu, W. GIS Partial Discharge Pattern Recognition Based on Time-Frequency Features and Improved Convolutional Neural Network. *Energies* **2022**, *15*, 7372. https://doi.org/10.3390/en15197372

Academic Editors: Guang Wang, Jiale Xie and Shunli Wang

Received: 24 August 2022
Accepted: 4 October 2022
Published: 7 October 2022

1. Introduction

The insulation state of gas-insulated switchgear (GIS) is closely related to the security of the power grid, and partial discharge (PD) is one of the critical indicators reflecting the internal insulation state of GIS [1,2]. The damage degree of insulation caused by different types of PD is quite different, so it is necessary to identify the PD signal in GIS to ensure GIS's safe and stable operation [3–5]. The feature extraction of PD is the key to affecting the recognition effect [6,7]. Currently, PD diagnosis methods mostly rely on statistics of characteristic parameters (e.g., phase, amplitude, and capacity of the discharge signal), and ignore the characteristic information of the discharge pulse itself. Therefore, the requirements for the statistical quantity of PD signal are relatively high, and discharge data of multiple power frequency cycles need to be counted [8]. The PD pulse itself carries rich feature information, and the features of different defects are different. Effective use of these features is of great significance to PD detection.

In recent years, scholars have conducted a lot of research on the feature extraction of PD pulse. In [9], PD time-domain waveform images were collected and converted into one-dimensional for pattern recognition. In [10], a feature extraction method combining wavelet packet analysis and singular value decomposition was adopted to extract features from frequency information. However, the pulse of PD is transient and unstable. Discharge occurs instantaneously, and the time and number of discharge pulses are random.

The time-varying information of PD signal is difficult to be represented by simple time-domain or frequency-domain analysis, and the performace of time-frequency joint analysis method is better than the conventional single domain analysis in diagnosing discharge characteristics [11]. As a time-frequency analysis method, wavelet transform can extract features from both time and frequency. By using its aspects of multi-resolution analysis, wavelet transform can characterize the local features of signals and provide more feature information for pattern recognition [12].

With the rapid development of deep learning, the convolutional neural network (CNN) has achieved good results in PD pattern recognition due to its powerful feature extraction ability. Compared with artificial neural networks and support vector machine, the training parameters are greatly reduced, and the recognition accuracy is improved [13–17]. In [18], a light-scale CNN was used to identify the simulated GIS PD data, which solved the problem of insufficient feature utilization. In [19], CNN and long and short-term memory networks were combined to improve the recognition accuracy by fusing the temporal and spatial features of PD signals. Considering the effect of CNN on PD recognition, this paper selects CNN to extract the features of the PD pulse. However, when the one-dimensional signal or the processed two-dimensional time-frequency diagram is used as the input alone, the local features of the PD pulse are seriously lost. A PD pattern recognition method based on dual-channel CNN is proposed to solve this problem, which fully utilizes the advantages of 1-D CNN and 2-D CNN.

This paper proposes an improved feature fusion convolutional neural network (IFCNN) model for PD pattern recognition. The time-domain features of the PD pulse are characterized by the feature sequence composed of discrete data points. The signal is subjected to wavelet transform, and the corresponding wavelet time-frequency diagram supplements the local features of the discharge transient signal. The CNN model with two parallel channels is used to extract the features of the time-domain feature sequence and wavelet time-frequency diagram. The local features of the discharge pulse in the time domain and frequency domain are mined through feature fusion. The dual-channel model is improved to avoid the loss of features in the training process. Specifically, a one-dimensional CNN is used to process the PD time-domain feature sequence, and a 1×1 convolutional layer is added after the convolutional layer to increase the nonlinearity of the one-dimensional model. A two-dimensional CNN is used to extract the features of the wavelet time-frequency diagram, and the shallow features are reduced in dimension and fused with the deep features. The improved dual-channel CNN model can automatically extract the time-frequency features of PD pulses in different dimensions. The main contributions of this paper are as follows:

(1) An IFCNN model is proposed for PD pattern recognition by extracting the features of PD pulses, which avoids counting a large number of discharge characteristic parameters. The improved model combines the advantages of 1-D CNN and 2-D CNN. It can automatically extract and fuse the features of the time-domain feature sequence and wavelet time-frequency diagram and obtain more feature information.
(2) The PD pulse signal is collected by experiment. At the same time, the mathematical model is used to simulate the PD source to obtain data to ensure the randomness of the signal and reflect the characteristics of the discharge pulse, avoiding the collection of a large amount of experimental data. The improved model is used to extract features automatically for training. The lightweight structure of the model makes it possible to apply in the field environment conveniently.
(3) The feasibility of the model is verified by experiments and compared with 1-D CNN and 2-D CNN of the same specification. The results show that the improved model has higher recognition accuracy by fusing the time-frequency features of different dimensions.

2. Proposed Method

2.1. Convolutional Neural Network

CNN is a feedforward neural network that can directly input the original image and avoid complex processing of the original signal. In recent years, it has been widely used in the field of pattern recognition [20–22]. CNN can extract the features of input data layer by layer and get the essential abstract representation of features. Figure 1 shows typical convolutional neural networks usually include an input layer, convolution layer, pooling layer, fully connected layer, and output layer. The convolution layer uses multiple convolution kernels to perform convolution calculations on the input data, extracts the corresponding data features, and then connects to the next layer through bias calculation and activation function. A mathematical formula can express the process:

$$X_i = \sigma(X_{i-1} \times W_i + b_i), \tag{1}$$

where X_i represents the output feature map of the ith layer, X_{i-1} represents the input feature map of the ith layer, W_i is the weight matrix of the ith convolution kernel, b_i is the offset vector of the ith layer, and *sigma* is the activation function. The main activation functions are *Tanh*, *Sigmoid*, and *ReLU*.

The pooling layer is generally divided into maximum pooling, mean pooling, and random pooling, which are used to reduce the output parameters of the convolution layer. The fully connected layer connects the features of the previous layer, and extracts and reduces the dimension of the features again. Finally, the output layer calculates the one-dimensional output sequence and obtains the probability value of each class to which the classification target belongs.

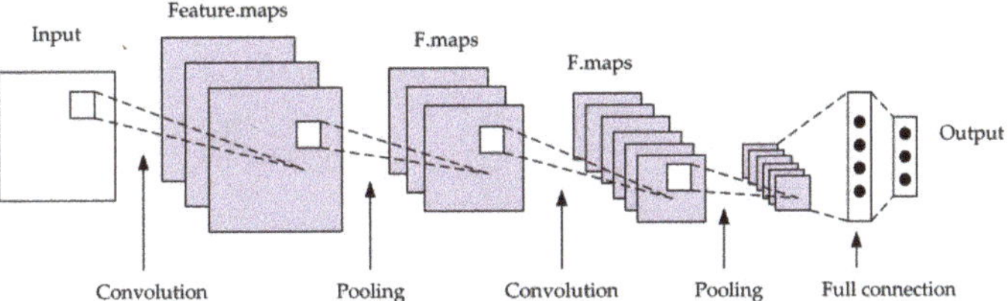

Figure 1. The structure of typical convolutional neural networks.

2.2. Improved Feature Fusion Convolutional Neural Network

In this paper, the IFCNN is used to extract the features of PD pulse to realize PD pattern recognition. The PD pulse waveform is a curve that changes the discharge intensity with time. The 1-D CNN can better express one-dimensional information of the PD pulse voltage, which is suitable for feature extraction of the time-domain waveform [9]. Therefore, the time-domain waveform features of PD pulse are transformed into one dimension, and 1-D CNN is used to extract the features. At the same time, the wavelet transform is applied to the signal, and the wavelet time-frequency diagram supplements the frequency-domain features of the PD signal extracted by the 2-D CNN. The specific model structure of IFCNN is shown in Figure 2. The model is composed of two parallel channels. The 1D-CNN inputs the one-dimensional time-domain feature sequence of the PD pulse, and the 2D-CNN inputs the two-dimensional wavelet time-frequency diagram. Both channels use convolutional and pooling layers alternately to extract features.

The 1D-CNN uses one-dimensional convolution and pooling kernels to adapt to the input one-dimensional time-domain feature sequence. It adds a 1×1 convolution layer after the ordinary convolution layer to access more activation functions and improve the nonlinear fitting ability of the one-dimensional network. The 2D-CNN uses a large convolution kernel in the first convolutional layer to increase the receptive field and obtain more features. In the feature extraction process, the shallow features pass through fewer convolution layers, the feature resolution is high, and the features contain more feature information, but the deep features have better semantics. To avoid the loss of features, the deep layer and shallow layer features are fused [23]. The traditional feature fusion directly characterizes the features of the two layers as feature vectors. It sends them to the fully connected layer, leading to too many parameters in the fully connected layer, resulting in a bloated model. Here, the output of the features by the first pooling layer of the network is passed through a 1 × 1 × 1 convolution layer to compress the feature channel to reduce the dimension of the data and retain the significant features, to realize the fusion of cross-channel features. The fully connected layer stretches the features of 1D-CNN and deep and shallow layers of 2D-CNN into feature vectors, and the feature vectors are spliced in the fusion layer to achieve feature fusion. The fully connected layer is used to continue the feature extraction and dimension reduction of the fused features. Finally, the Softmax classifier is used to to calculate the discharge type probability to achieve the PD classification.

In the IFCNN model, ReLU is selected as the activation function. To avoid gradient disappearance and explosion, the BN layer is used after the convolution layer to normalize the data and enhance the model's generalization ability. The pooling layer selects the maximum pooling to obtain the maximum value of the local area of the data to reduce the dimension. A fully connected layer is used after the fusion layer to prevent overfitting. The Dropout operation is used to randomly remove some neurons to solve the problem that the dimension of the feature vector increases after feature fusion.

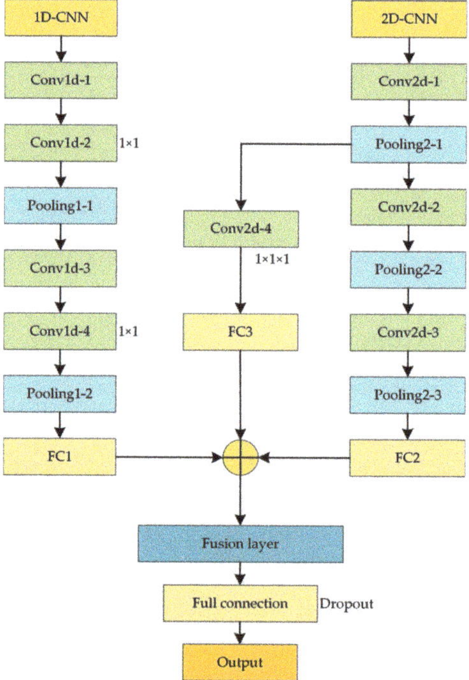

Figure 2. The structure of IFCNN.

3. PD Pattern Recognition

3.1. Data Acquisition

3.1.1. Experimental Data

Four typical PD defect models, including point discharge, surface discharge, air gap discharge and suspended discharge were selected for pattern recognition according to the possible defect types in field GIS. The experiment circuit is shown in Figure 3.

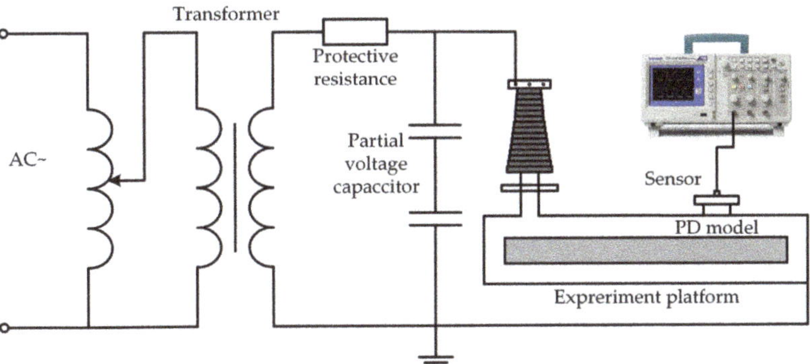

Figure 3. Partial discharge experiment circuit.

The power frequency high voltage control platform was used to apply high voltage to the PD defect model to generate PD signals. Figure 4 shows the experiment platform (Figure 4a) and the side view installation location (Figure 4b) of the defect. Four typical PD defect models are shown in Figure 5. The point discharge model simulated the presence of protrusions on the conductor surface in GIS. The surface discharge model simulated the existence of insulation defects on the surface of solid insulating materials in GIS. The air gap discharge model simulated the air gap inside the solid insulating material in GIS. The suspended discharge model simulated the poor contact of the conductor parts of GIS. When the conductor parts are energized, the potential difference between the potential suspension of the conductor parts and the surrounding parts will produce discharge.

The experiment adopted the method of stepwise pressurization, and the discharge pulse data were recorded and stored [24]. The sensor was a microwave antenna. A front RF amplifier was designed internally. Before signal transmission, it was amplified to improve the signal-to-noise ratio. The gain of the amplifier was 10 dB. A high pass filter was built in the channel to filter out interference signals. The detection frequency band of the sensor was 300 MHz to 1500 MHz, and the mean effective height was 9 mm. An oscilloscope was used to collect signals. The oscilloscope model was Tektronix7104 (bandwidth 1 GHz, maximum sampling rate 20 GSa/s). The oscilloscope sampling frequency was set to 10 MSa/s, and discrete data points were used to represent the waveform. The pulse waveforms of four kinds of PD signals are shown in Figure 6.

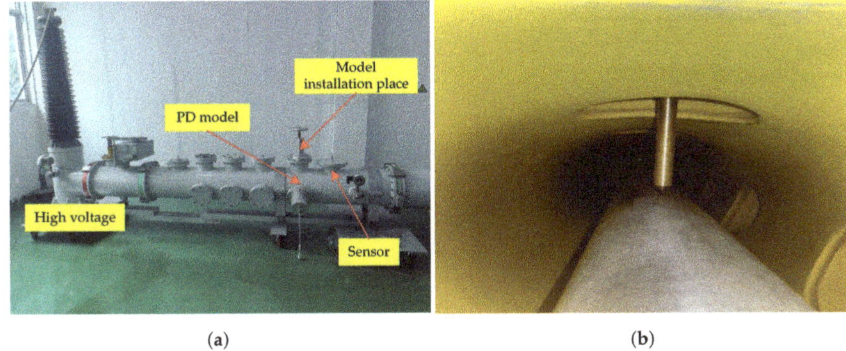

(a) (b)

Figure 4. PD experiment platform and inside the experiment platform: (**a**) PD experiment platform; (**b**) inside the experiment platform.

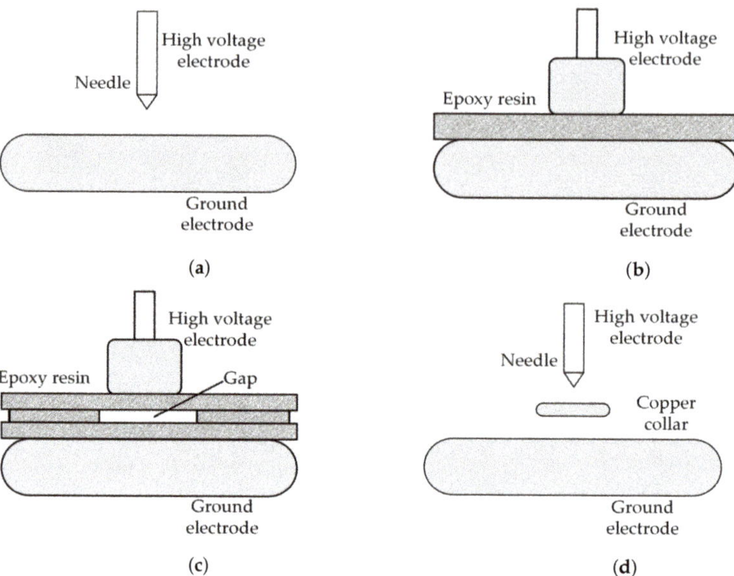

Figure 5. Typical PD defect models: (**a**) point discharge; (**b**) surface discharge; (**c**) air gap discharge; (**d**) suspended discharge.

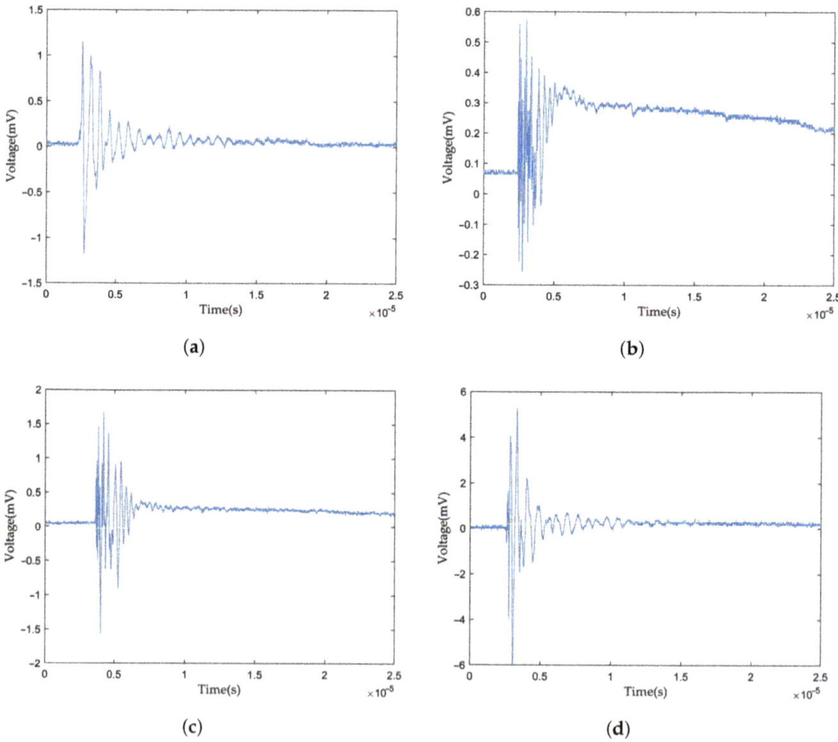

Figure 6. PD pulse waveforms: (**a**) point discharge; (**b**) surface discharge; (**c**) air gap discharge; (**d**) suspended discharge.

3.1.2. Simulation Data

To obtain as many PD fault samples as possible to reflect the characteristics of PD pulses and improve the accuracy of pattern recognition in this scheme, mathematical discharge models were used to simulate the PD source of typical PD defects to obtain the discharge data. Two mathematical models of single exponential decay oscillation pulse (SDOP) and double exponential decay oscillation pulse (DDOP) were used for simulation [25]. The expression can be expressed as:

$$\begin{cases} f_1(t) = Ae^{-t/\tau}\sin(2\pi ft) \\ f_2(t) = A(e^{-1.3t/\tau} - e^{-2.2t/\tau})\sin(2\pi ft), \end{cases} \tag{2}$$

where A is the signal amplitude, f is the signal oscillation frequency, and τ is the signal attenuation constant.

Both discharge models are established based on the $IEC60270$ measurement method, which is close to the actual PD signal obtained by the pulse current method, and can represent the signal collected in the project. The pulse waveforms of the two discharge mathematical models are shown in Figure 7.

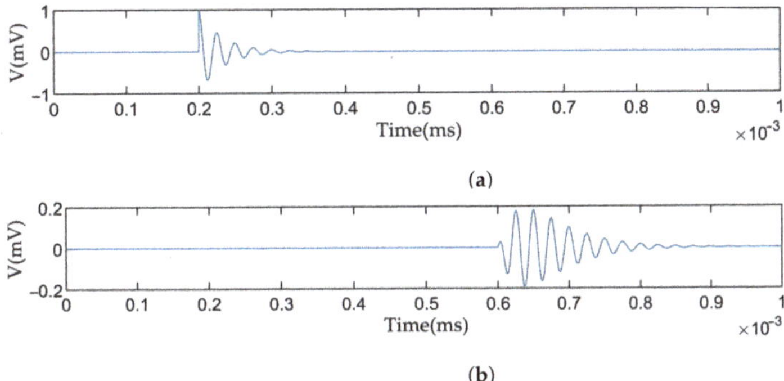

(b)

Figure 7. Pulse waveforms of discharge mathematical models: (**a**) SDOP; (**b**) DDOP.

The point , air gap, and suspended discharge were all superimposed by SDOP. The frequency of point discharge and air gap discharge is relatively low, the amplitude of the suspended discharge pulse is larger and the attenuation is more intense. DDOP superimposed the surface discharge. Referring to the PD simulation parameters in [26] and fitting the waveforms collected in the experiment, the statistical parameters were obtained to establish the PD source model. The specific parameters are shown in Table 1.

Table 1. PD simulation parameters.

Type	A/mV	τ/μs	f/KHz
Point discharge	0.3	0.5	120
Surface discharge	0.13	0.3	600
Air gap discharge	0.4	0.8	200
Suspended discharge	0.7	1.7	400

The simulated PD signals were collected in the simulated actual noise environment, and the simulated PD time-domain signals are shown in Figure 8.

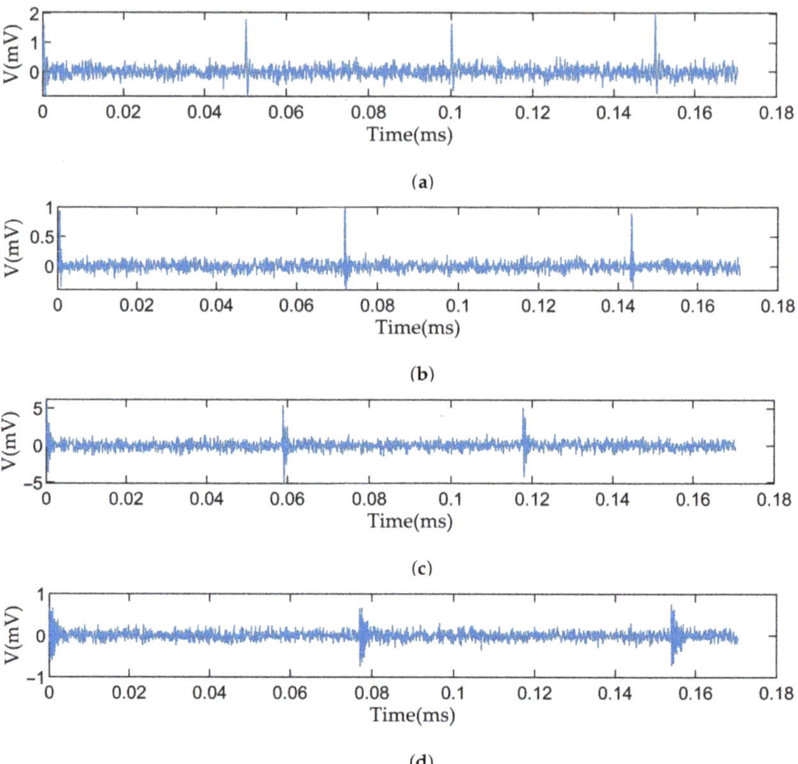

Figure 8. The simulated PD signals: (**a**) point discharge; (**b**) surface discharge; (**c**) air gap discharge; (**d**) suspended discharge.

3.2. Feature Extraction

3.2.1. Wavelet Transform

Wavelet transform is a local transform in the time domain and frequency domain. Wavelet transform performs multi-scale refinement analysis on the signal through scaling and translation operations, which can effectively extract the local features of the signal, and has a good effect on processing transient and non-stationary signals. In recent years, wavelet transform has often been used for de-noising PD signals [27,28], and has some applications in PD pattern recognition. For $f(t) \in L^2(R)$, its continuous wavelet transform can be expressed as:

$$WT(a,b) = \frac{1}{\sqrt{a}} \int_{-\infty}^{+\infty} f(t) \times \psi^*(\frac{t-b}{a})dt. \tag{3}$$

The basis function of wavelet transform is:

$$\psi_{a,b}(t) = \frac{1}{\sqrt{a}} \psi(\frac{t-b}{a}), \tag{4}$$

where $\psi(t)$ is the wavelet function. a is the scale factor, which is related to the frequency and controls the expansion and contraction of the wavelet function. b is the translation factor, which is related to time and controls the translation of the wavelet function. $\psi_{a,b}(t)$ is the result of scaling and shifting the wavelet function, and $\frac{1}{\sqrt{a}}$ is introduced for normalization.

Moving the wavelet in the time domain and taking the inner product with $f(t)$, the obtained wavelet coefficient reflects the similarity between the corresponding period signal and the current scale wavelet. The wavelet coefficients at different frequencies can be obtained by changing the frequency of the wavelet and stretching the length of the wavelet. Wavelet transform provides a window that changes with frequency. When dealing with high frequency, the time window is compressed and the time resolution is higher. Wavelet transform can adapt to the requirements of time-frequency analysis and select the resolution according to the characteristics of the signal. It can extract the local features of the PD signal. Compared with short-time Fourier transform, wavelet transform overcomes the problem that the window does not change with frequency. As long as the wavelet transform chooses a proper wavelet basis function, the time-frequency spectrum have a good focus and is more realistic than other transforms such as Hilbert-Huang.

3.2.2. Time-Frequency Analysis

For the PD pulses collected in the experiment, their time-domain features were represented by feature sequences composed of 2500 discrete sampling points. The complex Morlet wavelet was selected as the wavelet function. The PD pulse signals were subjected to wavelet transform, and the obtained wavelet time-frequency diagrams are shown in Figure 9.

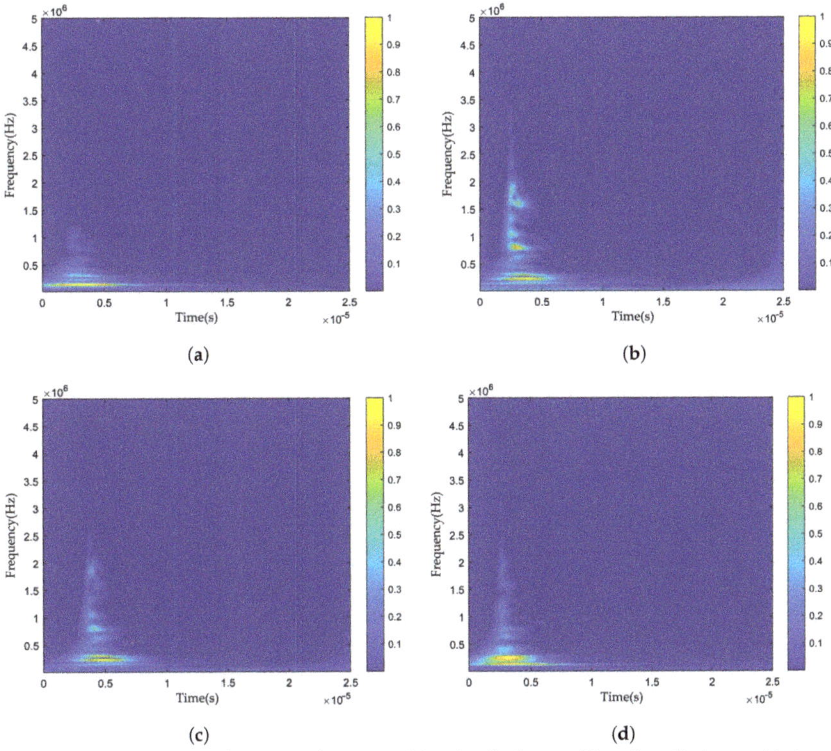

Figure 9. PD wavelet time-frequency diagrams: (**a**) point discharge; (**b**) surface discharge; (**c**) air gap discharge; (**d**) suspended discharge.

For the simulated PD time-domain signals, the wavelet threshold de-noising method was used for processing [29]. The signals were decomposed to the fifth layer using Sym3 wavelet function, and the signals were processed with a soft threshold to eliminate noise. The pulses of various simulated PD signals were effectively intercepted to facilitate the extraction of features for subsequent PD pattern recognition. The time-domain feature sequences were also used to describe the discharge pulse waveforms. Wavelet transform was performed on the intercepted PD pulse signals to obtain time-frequency features.

A complex Morlet wavelet was used for the wavelet transform to obtain wavelet time-frequency diagrams of various simulated PD pulses.

3.3. Model Training

Taking the time-domain feature sequence of PD pulse and the corresponding wavelet time-frequency diagram as a set of samples, 500 samples were selected for each of the four types of PD, including 200 experimental data and 300 simulated data. All samples were divided into a training set and validation set according to the ratio of 8:2. The data set was preprocessed, the discharge amplitudes of the time-domain feature sequences were normalized, the diagrams were grayed, and the corresponding labels were marked on the data set. The processed data set was input into the IFCNN model for supervised learning. The gradient descent algorithm was used in the training process, and the cross-entropy function was used as the loss function, which can be generally expressed as:

$$Loss = -\sum_{i=1}^{n} y_i \cdot \log y_i', \tag{5}$$

where y_i is the tag value, and y_i' is the predicted value.

Through iterative training, the weights between each layer were updated until the error reached the set expected value, the training was ended, and the trained model was saved. Finally, PD test data of unknown type was input into the trained model to verify the recognition effect of the model.

4. Results and Analysis

4.1. Training Process

The Pytorch framework based on Python3.6 was used to write the IFCNN model in this paper. The experimental hardware environment was an i7-6700HQ processor and 8G memory, and the software environment was the Windows10 operating system. The specific parameters of each layer of the IFCNN model built are shown in Table 2.

Table 2. The parameters of each layer of the IFCNN model.

Layer Number		Kernel Size	Step Size	Kernel Number
1D-CNN	Conv1d-1	1×5	1	6
	Conv1d-2	1×1	1	6
	Pooling1-1	1×3	3	6
	Conv1d-3	1×5	1	16
	Conv1d-4	1×1	1	16
	Pooling1-2	1×3	3	16
2D-CNN	Conv2d-1	5×5	1	6
	Pooling2-1	2×2	2	6
	Conv2d-2	3×3	1	16
	Pooling2-2	2×2	2	16
	Conv2d-3	3×3	1	32
	Pooling2-3	2×2	2	32
	Conv2d-4	1×1	1	1

The learning rate was set to 0.005, the number of iterations was set to 100, and the Batchsize was set to 64. The accuracy and loss curves of the training set and validation set in the training process are shown in Figure 10. As the number of iterations increases, the accuracy gradually increases and tends to be stable. The structure of the improved model is lightweight, and the training speed is fast. The model is suitable for PD pattern recognition.

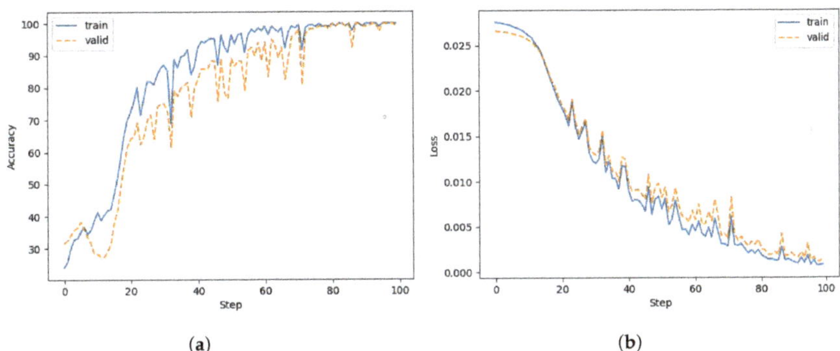

Figure 10. The accuracy and loss curves of IFCNN: (**a**) accuracy curve; (**b**) loss curve.

4.2. Accuracy Analysis of Pattern Recognition

The PD test data were inputted into the trained model for recognition, and the recognition accuracy P_r was used to evaluate the ability of PD recognition. The calculation P_r is:

$$P_r = \frac{N_r}{N_{sum}}, \tag{6}$$

where N_r is the number of samples whose identification type is consistent with the actual type, and N_{sum} is the total number of samples.

The confusion matrix of IFCNN pattern recognition results is shown in Figure 11, where 0, 1, 2 and 3 represent point discharge, surface discharge, air gap discharge and suspended discharge, respectively. It can be seen from the data in the figure that the IFCNN model has a high recognition accuracy, and the recognition accuracy of surface discharge reaches 98.3%. In order to ensure the stability of the model, the model was trained ten times, and the standard deviation of the overall recognition accuracy was 1.95.

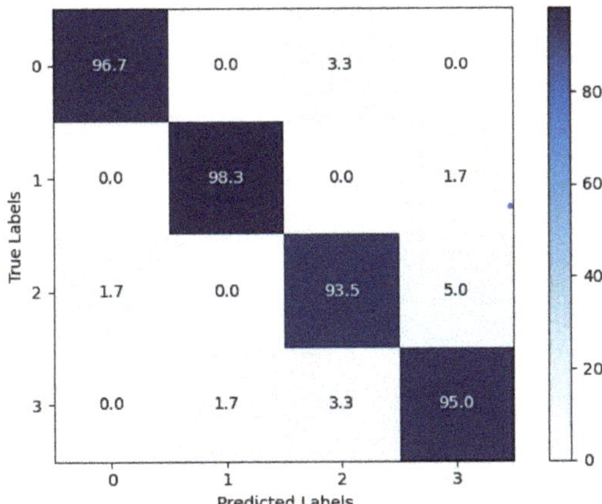

Figure 11. Confusion matrix of IFCNN.

In this paper, the experimental data and simulation data were combined with expanding the data set. In order to verify its impact on the experimental results, the experimental data and simulation data were separately used for testing. The identification results are shown in Table 3. The results show that the improved models have a good recognition effect.

Table 3. PD recognition results for different data sets.

P_r	**Point Discharge**	**Surface Discharge**	**Air Gap Discharge**	**Suspended Discharge**
Experimental data	95.8%	98.3%	95%	96.7%
Simulation data	99.2%	96.7%	97.5%	97.5%
Mixed data	96.7%	98.3%	93.3%	95%

The quality of wavelet transform feature extraction depends on the similarity between the wavelet waveform and the measured signal waveform. Different wavelet functions were used to verify the influence of different wavelet functions on pattern recognition accuracy. The appropriate center frequency was selected to perform wavelet transform on PD signals. The wavelet functions commonly used for signal processing include Bior, Sym, Db, and Morlet. The processed data sets were respectively input into the IFCNN model for testing. The pattern recognition results are shown in Table 4.

Table 4. PD recognition results in different wavelet functions.

P_r	**Bior**	**Sym**	**Db**	**Morlet**
Overall accuracy	91.7%	89.6%	93.3%	95.8%

The results in Table 4 show that the recognition accuracy is higher using the complex Morlet wavelet. The complex Morlet wavelet function is a complex-valued function multiplied by a Gaussian function and a complex trigonometric function, and its waveform characteristics are more similar to the PD pulse waveform. The complex Morlet wavelet function can reflect the time-domain and frequency-domain features of the discharge signal and obtain the distribution of signal energy with time and frequency.

4.3. Comparison of Different Methods

To verify the effect of the proposed method on the feature extraction and classification of PD pulses, the time-domain feature sequences, and wavelet time-frequency diagrams were input into the 1D-CNN and 2D-CNN of the same specification for comparative experiments. The recognition accuracy was compared using the same data set. The pattern recognition results of different methods are shown in Table 5.

Table 5. PD recognition results from different methods.

P_r	**Feature Sequence + 1D-CNN**	**Wavelet Diagram + 2D-CNN**	**IFCNN**
Point discharge	78.3%	88.3%	96.7%
Surface discharge	81.7%	90%	98.3%
Air gap discharge	75%	91.7%	93.3%
Suspended discharge	78.3%	86.7%	95%
Overall accuracy	78.3%	89.2%	95.8%

It can be seen from the data in Table 5 that the recognition accuracy of the IFCNN has reached 95.8%, and the recognition accuracy is much higher than that of the single-channel CNN. Due to the similarity of the pulse waveforms of air gap discharge and suspended discharge, the accuracy of pattern recognition is low when only the time-domain feature sequences are used to extract the time-domain features of PD pulses. The wavelet time-frequency diagrams extract features from both the time and frequency domains, improving recognition accuracy. The improved algorithm in this paper fuses the two features and avoids the problem of insufficient feature utilization through structural improvement, which further improves the recognition accuracy. During the training process, the improved model converges faster under the same number of iterations.

5. Conclusions

In this paper, an IFCNN model is constructed to extract the features of PD pulses, and PD pattern recognition is realized by using the features of the pulse signal in the time domain and frequency domain, which solves the problem that traditional detection methods need a large amount of statistical discharge data. The improved model takes the one-dimensional time-domain feature sequence of PD pulse and wavelet time-frequency diagram as input signals, uses the two-channel CNN to extract the features, fuses the extracted fault information, and finally uses the Softmax layer to realize the classification of PD. The method combines the advantages of 1D-CNN and 2D-CNN, fuses the time-frequency features of different dimensions, and mines more feature information.

The data set was established to train and test the models by establishing four typical PD defect models and using mathematical models to obtain the discharge pulse data of different PD types. Compared to the pattern recognition effect of the improved model with 1D-CNN and 2D-CNN, the overall recognition rate of the IFCNN model reaches 95.8%, followed by 2D-CNN (89.2%) and 1D-CNN (78.3%). The recognition effect of IFCNN is higher than the traditional single-channel model, due to the reason that can fully extract the time-frequency features of the discharge pulse and further retain the feature information through structural optimization. In the actual field environment, different sensors and measuring circuits may affect the features of the collected discharge pulses. In order to ensure the recognition accuracy of the improved algorithm, the algorithm can be trained by re-collecting the discharge pulse data, then use the features of the extracted discharge pulse to realize PD recognition. The improved algorithm extracts the features of the attenuation period of the single pulse. It is unnecessary for the collector to collect the discharge data of multiple power frequency cycles, leading the little storage space for hardware devices. The structure of the improved algorithm is lightweight, the number of convolution layers and the requirements for the operating system are small are small, the model can be recognized when it is transplanted to the embedded system, and also be easily applied to the field environment.

Author Contributions: Conceptualization, J.Z. and Z.C.; methodology, J.Z., Z.C. and Q.W.; software, Z.C.; validation, J.Z., Z.C. and W.X.; formal analysis, Z.C. and W.X.; investigation, Z.C. and Q.W.; resources, J.Z. and H.Q.; data curation, J.Z., Z.C. and Q.W.; writing—original draft preparation, Z.C.; writing—review and editing, J.Z., Z.C. and W.X.; visualization, Z.C.; supervision, J.Z. and H.Q. All authors have read and agreed to the published version of the manuscript.

Funding: This research was funded by the Postgraduate Research & Practice Innovation Program of Jiangsu Province under the grant number [SJCX21_1283].

Conflicts of Interest: The authors declare no conflict of interest.

References

1. Thi, N.; Do, T.D.; Jung, J.R.; Jo, H.; Kim, Y.H. Anomaly Detection for Partial Discharge in Gas-Insulated Switchgears Using Autoencoder. *IEEE Access* **2020**, *8*, 152248–152257. [CrossRef]
2. Khan, Q.; Refaat, S.S.; Abu-Rub, H.; Toliyat, H.A. Partial Discharge Detection and Diagnosis in Gas Insulated Switchgear: State of the Art. *IEEE Electr. Insul. Mag.* **2019**, *35*, 16–33. [CrossRef]
3. Jing, Q.; Yan, J.; Lu, L.; Xu, Y.; Yang, F. A Novel Method for Pattern Recognition of GIS Partial Discharge via Multi-Information Ensemble Learning. *Entropy* **2022**, *24*, 954. [CrossRef] [PubMed]
4. Gao, A.; Zhu, Y.; Cai, W.; Zhang, Y. Pattern Recognition of Partial Discharge Based on VMD-CWD Spectrum and Optimized CNN With Cross-Layer Feature Fusion. *IEEE Access* **2020**, *8*, 151296–151306. [CrossRef]
5. Song, S.; Qian, Y.;Wang, H.; Zang, Y.; Sheng, G.; Jiang, X. Partial Discharge Pattern Recognition Based on 3D Graphs of Phase Resolved Pulse Sequence. *Energies* **2020**, *13*, 4103. [CrossRef]
6. Haikun, S.; Kwok, L.; Feng, L. Partial Discharge Feature Extraction Based on Ensemble Empirical Mode Decomposition and Sample Entropy. *Entropy* **2017**, *19*, 439.
7. Wang, K.; Li, J.; Zhang, S.; Gao, F.; Cheng, H.; Liu, R.; Liao, R.; Grzybowski, S. A New Image-Oriented Feature Extraction Method for Partial Discharges. *IEEE Trans. Dielectr. Electr. Insul.* **2015**, *22*, 1015–1024. [CrossRef]
8. Xu, C.; Chen, J.; Liu, W.; Lv, Z.; Li, P.; Zhu, M. Pattern Recognition of Partial Discharge PRPD Spectrum in GIS Based on Deep Residual Network. *High Volt. Eng.* **2022**, *48*, 1113–1123.

9. Wan, X.; Song, H.; Luo, L.; Li Z.; Sheng, G.; Jiang X. Application of Convolutional Neural Networks in Pattern Recognition of Partial Discharge Image. *Power Syst. Technol.* **2019**, *43*, 2219–2226.
10. Zhao, X.; Liu, X.; Meng, Y.; Liu, S.; Chai, Q.; Wu, P.; Meng, G. Partial Discharge Pattern Classification by Singular Value Decomposition of Wavelet Packet Energy Features. *J. Xi'An Jiaotong Univ.* **2017**, *51*, 116–121.
11. Ren, Z. Identification of Gabor Distribution Characteristics of Partial Discharge on Typical SF6 Gas Insulated Defects. *High Volt. Appar.* **2015**, *51*, 142–148.
12. Li, P.; Tian, Q.; Huo, M.; Chen, X.; Lin, Y.; Li, J. Recognition Method of Partial Discharge Type of Transformer Based on Wavelet Transform and Hog Feature. *Electr. Drive* **2021**, *51*, 52–56.
13. Liu, B.; Zheng, J. Partial Discharge Pattern Recognition in Power Transformers Based on Convolutional Neural Networks. *High Volt. Appar.* **2017**, *53*, 70–74.
14. Song, H.; Dai, J.; Sheng, G.; Jiang, X. GIS Partial Discharge Pattern Recognition via Deep Convolutional Neural Network under Complex Data Source. *IEEE Trans. Dielectr. Electr. Insul.* **2018**, *25*, 678–685. [CrossRef]
15. Wang, Y.; Yan, J.; Yang, Z.; Wang, J.; Geng, Y. A Novel 1DCNN and Domain Adversarial Transfer Strategy for Small Sample GIS Partial Discharge Pattern Recognition. *Meas. Sci. Technol.* **2021**, *32*, 125118. [CrossRef]
16. Huang, X.; Xiong, J.; Zhang, Y.; Liang, J. Partial Discharge Pattern Recognition of Switchgear Based on Residual Convolutional Neural Network. *J. Phys. Conf. Ser.* **2020**, *1659*, 012057. [CrossRef]
17. Florkowski, M. Classification of Partial Discharge Images Using Deep Convolutional Neural Networks. *Energies* **2020**, *13*, 5496. [CrossRef]
18. Wang, Y.; Yan, J.; Yang, Z.; Liu, T.; Zhao, Y.; Li, J. Partial Discharge Pattern Recognition of Gas-Insulated Switchgear via a Light-Scale Convolutional Neural Network. *Energies* **2019**, *12*, 4674. [CrossRef]
19. Liu, T.; Yan, J.; Wang, Y.; Xu, Y.; Zhao, Y. GIS Partial Discharge Pattern Recognition Based on a Novel Convolutional Neural Networks and Long Short-Term Memory. *Entropy* **2021**, *23*, 774. [CrossRef] [PubMed]
20. Wang, Y.; Yan, J.; Sun, Q.; Li, J.; Yang, Z. A MobileNets Convolutional Neural Network for GIS Partial Discharge Pattern Recognition in the Ubiquitous Power Internet of Things Context: Optimization, Comparison, and Application. *IEEE Access* **2019**, *7*, 150226–150236. [CrossRef]
21. Peng, X.; Yang, F.; Wang, G.; Wu, Y.; Li, L.; Li, Z. A Convolutional Neural Network-Based Deep Learning Methodology for Recognition of Partial Discharge Patterns from High-Voltage Cables. *IEEE Trans. Power Deliv.* **2019**, *34*, 1460–1469. [CrossRef]
22. Sun, Y.; Ma, S.; Sun, S.; Liu, P.; Zhang, L.; Ouyang, J.; Ni, X. Partial Discharge Pattern Recognition of Transformers Based on MobileNets Convolutional Neural Network. *Appl. Sci.* **2021**, *11*, 6984. [CrossRef]
23. Li, H.; Zhu, Y.; Wang, J. Transformer PRPD Pattern Recognition Based on Multi-layer Feature Fusion CNN. *Electr. Meas. Instrum.* **2020**, *57*, 63–68.
24. Nasr, E.A.; Shahabi, S.; Kordi, B. Partial Discharge Detection and Identification at Low Air Pressure in Noisy Environment. *High Volt.* **2021**, *6*, 850–860. [CrossRef]
25. Xiao, S.; Chen, B.; Shen, D.; Chen, H. Application of Improved VMD and Threshold Algorithm in Partial Discharge Denoising. *J. Electron. Meas. Instrum.* **2021**, *35*, 206–214.
26. Zhang, M. Study on Partial Discharge Feature Extraction and Pattern Recognition of High Voltage Switchgear. Master's Thesis, Shanghai DianJi University, Shanghai, China, 2020.
27. Hussein, R.; Shaban, K.B.; El-Hag, A.H. Wavelet Transform With Histogram-Based Threshold Estimation for Online Partial Discharge Signal Denoising. *IEEE Trans. Instrum. Meas.* **2015**, *64*, 3601–3614. [CrossRef]
28. Jian, L. Wavelet De-noising of Partial Discharge Signals Based on Genetic Adaptive Threshold Estimation. *IEEE Trans. Dielectr. Electr. Insul.* **2012**, *19*, 543–549.
29. Mi, H.; Wang, X.; Ren, G.; Zhen, Y. De-noising for White Noise in Partial Discharge Signals by Adaptive Wavelet Threshold Estimation. *High Volt. Appar.* **2021**, *57*, 94–101.

Article

An Efficient Method Combined Data-Driven for Detecting Electricity Theft with Stacking Structure Based on Grey Relation Analysis

Rui Xia, Yunpeng Gao *, Yanqing Zhu, Dexi Gu and Jiangzhao Wang

College of Electrical and Information Engineering, Hunan University, Changsha 410082, China
* Correspondence: gaoyp@hnu.edu.cn; Tel.: +86-136-0731-9138

Abstract: Nowadays, electricity theft has been a major problem worldwide. Although many single-classification algorithms or an ensemble of single learners (i.e., homogeneous ensemble learning) have proven able to automatically identify suspicious customers in recent years, after the accuracy of these methods reaches a certain level, it still cannot be improved even if it continues to be optimized. To break through this bottleneck, a heterogeneous ensemble learning method with stacking integrated structure of different strong individual learners for detection of electricity theft is presented in this paper. Firstly, we use the grey relation analysis (GRA) method to select the heterogeneous strong classifier combination of LG + LSTM + KNN as the base model layer of stacking structure based on the principle of the highest comprehensive evaluation index value. Secondly, the support vector machine (SVM) model with relatively good results of the stacking overall structure experiment is selected as the model of the meta-model layer. In this way, a heterogeneous integrated learning model for electricity theft detection of the stacking structure is constructed. Finally, the experiments of this model are conducted on electricity consumption data from State Grid Corporation of China, and the results show that the detection performance of the proposed method is better than that of the existing state-of-the-art detection method (where the area under receiver operating characteristic curve (AUC) value is 0.98675).

Keywords: electricity theft; stacking structure; analytic hierarchy process; entropy weight method; grey relation analysis

Citation: Xia, R.; Gao, Y.; Zhu, Y.; Gu, D.; Wang, J. An Efficient Method Combined Data-Driven for Detecting Electricity Theft with Stacking Structure Based on Grey Relation Analysis. *Energies* **2022**, *15*, 7423. https://doi.org/10.3390/en15197423

Academic Editor: Anastasios Dounis

Received: 1 September 2022
Accepted: 7 October 2022
Published: 10 October 2022

1. Introduction

Electricity theft in the power system refers to malicious users tampering with electricity meters or attacking smart grids through a specific technology or devices in order to reduce or not pay electricity bills. Electricity theft seriously damages the economic interests of power companies, and the direct economic loss of State Grid Corporation of China due to electricity theft exceeds 1 billion yuan each year [1]. In January 2017, a research report released by the Northeast Group, a power grid consulting firm, said that the annual economic losses caused by non-technical losses in the 50 developing countries surveyed by it totaled $64.7 billion [2]. The worst of them is in India. India's annual revenue loss caused by electricity theft amounts to $17 billion US dollars [3]. Neither is this solely an issue in developing countries: relatively large revenue losses caused by electricity theft occur in developed countries as well, e.g., the revenue losses from electricity theft in the United Kingdom and the United States are as high as $6 billion per year [4]. At the same time, theft of electricity poses a huge threat to the order of market electricity consumption and the stable operation of the power grid. In areas where electricity theft is common (such as India), the power consumption side encounters irregular voltage dips and intermittent power interruptions, especially during peak loads, which can cause fires and threaten community safety in severe cases [5]. Therefore, it is necessary to accurately detect the behavior of electricity theft and provide technical support for the grid company to further identify the users suspected of electricity theft.

The existing electricity stealing methods mainly include the undervoltage method, the undercurrent method, the phase shift method, the differential expansion method, and the no-table method in terms of physical means. The above physical methods can be roughly divided into three categories, as shown in Figure 1 for the three categories of methods [1], respectively.

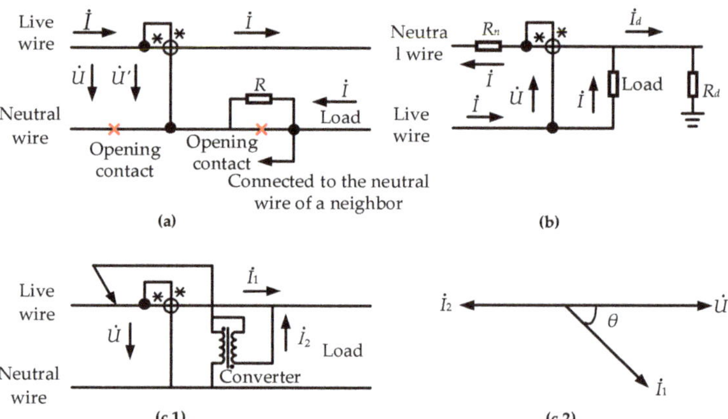

Figure 1. Three types of physical electricity theft methods. (**a**) Voltage reduction type wiring diagram. (**b**) Current reducing type wiring diagram. (**c.1**) Power factor reduction type wiring diagram. (**c.2**) Power factor reduction type phase diagram.

Figure 1a is a voltage reduction type. The unlawful user disconnects the zero line terminal and then connects it to the neighbor's zero line through the large resistance R. The electric energy meter is connected in series with the large resistance R to divide the voltage, and the electric energy meter measures the voltage $U' = R_1/(R_1 + R) \times U$, where R_1 is the resistance of the electric energy meter, U is the actual voltage, and the electric energy meter only measures the voltage obtained by its partial pressure, which reduces the measured electricity consumption. This is supposed to make the electric energy meter lose voltage or the measured voltage to be lower than the actual voltage by operating the voltage measurement loop, which indirectly causes the electricity consumption measured by the electric energy meter to decrease or be zero, thereby realizing electricity stealing.

Figure 1b is the current reducing type, where R_n is the zero line impedance, R_d is the grounding impedance, and I is the load current. The unlawful user will ground the neutral line after swapping the neutral line and the live line, and shunt R_n and R_d in parallel, so that the flow through the current of the electric energy meter $I_0 = R_d/(R_n + R_d) \times I$. The electric energy meter only measures the current divided by R_n, which reduces the measured electricity consumption. The current measured by the electric energy meter is zero or lower than the actual current by operating the current measurement loop. The electric current indirectly causes the electricity consumption measured by the electric energy meter to be reduced or zero, thereby realizing electricity theft.

Figure 1c.1 is the type that reduces the power factor. The unscrupulous user connects the modified specific converter to the circuit in parallel, so that the current flowing through the energy meter is the vector sum of the load current I_1 and the converter current I_2. The current flowing into the electric meter in the same phase as the voltage is $I_1 cos\theta - I_2$ makes the electric energy meter rotate slowly, stop, or reverse with the change of the size and nature of the load. By increasing the phase difference between the current and the voltage, the power factor measured by the electric energy meter decreases or becomes negative, which indirectly causes the electricity consumption measured by the electric energy meter to be reduced, zero, or negative, thereby realizing electricity stealing. The phase representation is shown in Figure 1c.2.

With the intelligent development of science and technology, many high-tech power stealing methods continue to emerge. For example, some unscrupulous users install remote control devices inside and outside the electric energy meter, and then intelligently control the on-off of the circuit and the size of the series-inserted resistance equipment. The timer outputs the neutral point intermittently. With the popularization of time-of-use electricity prices, some lawbreakers achieve the purpose of stealing electricity by reversing the timing of electricity consumption.

The traditional electricity stealing detection method requires manual on-site investigation, which is labor-intensive and has low detection efficiency and high blindness. At present, some experts and scholars have developed anti-theft devices based on the mechanism of electricity theft, which can effectively prevent the occurrence of certain electricity theft behaviors [6,7]. However, since it is only designed for some traditional electricity stealing means or some new types of electricity stealing means, the universality of the anti-electricity stealing device is low, and at the same time, the hardware cost and the possibility of hardware failure are increased. With the continuous improvement of power grid intelligence, power companies have obtained massive power consumption data to provide strong support for data mining methods. Based on data mining methods, the implicit information behind the data can be obtained. How to effectively use power big data to achieve efficient and accurate anti-theft malicious user identification has become particularly important.

1.1. Literature Review

Electricity theft detection methods based on data mining can be mainly divided into three categories. The first category is to realize electricity theft detection by building statistical models to analyze network status information such as grid voltage, current, power and network topology [8–11]. The electricity theft detection method based on the statistical model needs to obtain the grid network topology, network parameters, and the correct household change relationship. Due to the complex and dynamic change of the power grid network structure, this method has great limitations in practical engineering applications.

The second category is the game theory detection method. From the perspective of economics, this method builds a game theory model between power supply enterprises and electricity malicious users to quantify the benefits of electricity theft and governance [12–14]. For example, in [12], a Stackelberg game theory model was established to analyze the strategic interaction between a power company and multiple electricity malicious users, and the sampling rate and threshold were tested for likelihood ratios according to the Stackelberg equilibrium. Another example is the intrusion defense model based on game theory in [14], which combines honeypot technology with game theory, and obtains the optimal strategy for both sides of the attack through the game tree. Although the above game theory method has well described the interest relationship between power supply enterprises and electricity malicious users, the current research on the detection method of electricity theft based on game theory mainly stays at the level of theoretical derivation and simulation, which is temporarily difficult to apply to engineering practice.

The third category is the construction of electricity theft detection model based on data-driven method mining of electricity data information. Data-driven methods can be divided into unsupervised learning, semi-supervised learning and supervised learning according to the amount of prior knowledge required. Among them, unsupervised learning can automatically extract the typical characteristics of users' electricity consumption by learning the inherent similar correlation attributes of user electricity consumption data, cluster normal users, and find outliers as abnormal users [15]. In [16], the authors proposed an electricity stealing detection model based on cluster point algorithm, but because there is no feature extraction process and the algorithm is simple, the detection accuracy is low. In [17], the authors proposed feature extraction based on time-scale load sequence and constructed a sequential ensemble detector based on a deep auto-encoder with attention (AEA), gated recurrent units (GRUs), and feed forward neural networks to detect electricity theft behavior, but the feature extraction process is complex and computationally expensive. Reference [18] proposes a generative adversarial network to generate realistic electricity stealing samples, enhance the diversity of electricity stealing samples, and simplify the modelling process. However, the unsupervised learning method relies heavily on parameters and is not suitable for complex power grid environments and the detection of various types of electricity theft methods.

The semi-supervised learning method uses a small amount of label data obtained to train the initial learner, test and classify the unknown category data, and add the samples with high confidence coefficient in the classification results to the training set to train the model again, and repeat this process until all samples are the most excellent classification. Reference [19] uses a correlation denoising autoencoder to achieve feature extraction and feature association of electricity data. In [20], the authors propose a semi-supervised learning-based SSAE generation model and design an adversarial module to enhance the model's anti-noise ability. In [21], the authors adopted a semi-supervised learning method based on consistency loss to solve the problem of less label data in electricity stealing detection. There is a serious data imbalance problem in electricity stealing detection. There are fewer known labels in a small number of electricity stealing samples, which is easy to cause overfitting of the semi-supervised model and cannot effectively identify other types of anomalies. The method requires part of the label information, so the quality of the initial label data is high, and semi-supervised learning needs to solve the problems of overfitting and high-quality labels. Therefore, in the actual power grid situation, the applicability of this type of method is not high.

In order to overcome the shortcomings of unsupervised learning methods and semi-supervised learning methods for electricity theft detection, supervised learning methods can be used to detect electricity theft. The supervised learning method requires part of the label data confirming the user steals electricity as a training set, and uses the trained model to test and classify the unknown category data. Supervised learning learns the implicit information in the feature quantity according to the label information, finds the relationship between the feature quantity and the label information, and detects the unknown category data according to it. When using SVM or decision tree method, if the power consumption data set contains noise, such as missing data, the detection performance is poor [22,23]. For the high-dimensional data of user power consumption, the detection model of shallow structure cannot effectively process it. In order to further improve the detection accuracy, ensemble learning methods such as XGboost are applied in the field of electricity stealing detection [24,25]. However, the above methods do not perform feature extraction on the data, cannot find the time series features of electricity consumption data, and cannot achieve accurate prediction and classification when dealing with massive electricity consumption data. To solve the feature extraction problem, a new feature-engineering framework for theft detection in smart grids is introduced, however this method is complex and computationally intensive [26]. For this purpose, neural networks [27] and LSTM [28] can be used for feature extraction and classification prediction. However, because neural networks or their variants are prone to overfitting due to excessive network training times and long model training time, in addition, it is difficult to optimally set the hyperparameters, which leads to the detection accuracy reaching a certain level, which cannot be improved even if the optimization is continued.

1.2. Motivation

In order to break through the bottleneck of the existing single-classification algorithm or fusion algorithm, when the accuracy of electricity theft behavior detection reaches a certain level, even if it continues to optimize, it still cannot be improved [29,30]. For their optimization algorithms, such as the stacking strong model ensemble learning method, the selection of base classifiers does not have a good selection strategy, resulting in poor detection results or unable to explain the rationality of its selection. Moreover, these optimization methods do not take into account the complexity of the model [31]. In this paper, we use a multi-model fusion integrated learning algorithm based on the stacking structure to address the above problems.

The main contributions of this paper are summarized as follows:

1. This paper considers that while improving the accuracy and generalization ability of stacking structure algorithm and reducing the complexity of the model, the combined weight method of subjective weight and objective weight based on grey relation analysis (GRA) [32] is used to determine the weight of a single performance index of the classifier.
2. We extract the user's effective features of electricity consumption through a statistical-based method and reduce the dimensionality of the extracted features using the principal component analysis (PCA) method to reduce the redundancy of the data.
3. For the stacking structure, the choice of the base model is a difficult problem for all researchers. We conducted a large number of experiments and compared and analyzed the combination experiments of different models, and obtained the base model combination with excellent detection results and model complexity. In addition, for our chosen meta-model, SVM, we use particle swarm optimization (PSO) to optimize its parameters to get a better detection result.

The remainder of the paper is structured as follows. Data preparation is introduced in Section 2, which includes the recovery of missing values in the original dataset and the repair of outliers, as well as feature extraction and dimensionality reduction of the dataset. The stacking integrated structure is described in Section 3. Numerical experiments are conducted, and the analysis of experiments results is shown in Section 4. Final remarks are then presented in Section 5.

2. Data Preparation

In this section, the preprocessing process method based on the original dataset, including the interpolation of missing values and the repair of outliers, is introduced in detail. The feature extraction of electricity consumption dataset is then described.

2.1. Dataset

The dataset is gathered from smart meters of electricity consumption and was obtained from a province of the State Grid Corporation of China. The dataset is a sequence of daily electricity consumption, which is characterized as a time series, and records the daily electricity consumption of 9956 users from 1 January 2015 to 31 December 2015. The data are divided into thieves and normal electrical consumers, where the thieving consumers compose 14% of the total. The dataset description is shown in Table 1 [27].

Table 1. The Description of Dataset.

Timeline	Number of Normal Customers	Number of Theft Customers	The Total Number of Customers
2015/01/01–2015/12/31	8562 (86%)	1394 (14%)	9956 (100%)

2.2. Data Preprocess

In the process of collecting electricity load data, due to software and hardware failures, special events, and other factors, the data may contain missing or some erroneous values, which will affect the continuity of electricity consumption records, so it is necessary to process the original dataset.

This paper uses the method named "three-sigma rule of thumb" to recover the missing values [27], and the formula is as follows:

$$f(x_i) = \begin{cases} \frac{x_{i-1}+x_{i+1}}{2} \text{ if } x_i > 3 \cdot \sigma(\mathbf{x}_i) \text{ and } x_{i-1}, x_{i+1} \notin \text{NaN} \\ 0 \quad x_i \in \text{NaN}, x_{i-1} \text{ or } x_{i+1} \notin \text{NaN} \\ x_i \quad\quad x_i \notin \text{NaN} \end{cases}, \tag{1}$$

where x_i represents the power consumption value of a user in a day, $\sigma(\mathbf{x}_i)$ represents the standard deviation of vector $\mathbf{x}_i$, denote NaN as if x_i is not a number value.

In addition, for the outliers in the dataset, the following formula is used to recover [27]:

$$f(x_i)\begin{cases} \text{mean}(\mathbf{x}_i) & \text{if } x_i \in \text{NaN} \\ x_i & \text{others} \end{cases}, \tag{2}$$

where mean($\mathbf{x}_i$) represents the average of vector $\mathbf{x}_i$.

The power consumption habits of each power user are different. If the load data is not standardized, some users with high power consumption levels will have a greater impact on the detection model, which will increase the burden of the algorithm and is not conducive to model training. Extreme cases may lead to the model struggling to converge. Data standardization can be performed using some mathematical transformation processing to convert the original data to a fixed value range. The power load includes base load and variable load. The use of min-max standardization can remove the base load and highlight the trend of the variable load, while avoiding the impact of large differences in orders of magnitude. The daily load can be normalized to reduce the abnormal number of days and seasonal effects with critical peaks or false data injection. The min-max standardized calculation formula [25] is:

$$x^k_{i,j} = \frac{x^k_{i,j} - x^k_{imin}}{x^k_{imax} - x^k_{imin}}, \tag{3}$$

where x^k_{imin} is the minimum value of the kth day load for the ith user, and x^k_{imax} is the maximum value of the kth day load for the ith user.

2.3. Feature Extraction

Through the full understanding and comprehensive analysis of the user electricity dataset, it can be seen that there are certain differences in the fluctuations and trends of the electricity load between normal users and electricity users [33], and after extracting valuable information about the user electricity consumption data, the established model can be made to more accurately reflect the difference between the data and obtain better training results. Statistics are extracted from the after-preparation electricity consumption sequence as time series features, which are characterized by D_1–D_{49}, and the statistics-based features are shown in Table 2.

Table 2. The characteristic indicators of user electricity consumption time series statistical.

Characteristic Indicators	Dimension
Standard deviation and discrete coefficient of annual electricity consumption	D_1, D_2
Standard deviation and discrete coefficient of quarterly electricity consumption	D_3~D_6, D_7~D_{10}
Standard deviation and discrete coefficient of monthly electricity consumption	D_{11}~D_{21}, D_{22}~D_{32}
Average monthly electricity consumption rising and falling trends	D_{33}~D_{41}
The maximum and minimum value of the difference and the ratio of the average electricity consumption in the adjacent two months	D_{42}~D_{43}, D_{44}~D_{45}
The maximum and minimum value of the difference and the ratio of the average electricity consumption in the adjacent two quarters	D_{46}~D_{47}, D_{48}~D_{49}

Note the user electricity data set $X = (x_n, n = 1, 2, \ldots, N)$ after preprocessing, where N is the number of users. The user's daily electricity consumption sequence is $x_n = \{x_{nd}, d = 1, 2, \ldots, D\}$, the monthly electricity consumption sequence is $y_n = \{y_{nm}, m = 1, 2, \ldots, M\}$, the quarter power consumption sequence is $z_n = \{z_{nq}, q = 1, 2, \ldots, Q\}$, where the user's electricity consumption time is collected is D days, M months, Q quarters. The standard deviation of electricity consumption is std, which indicates the fluctuating characteristics of electricity consumption data [33]:

$$std = \sqrt{\frac{\sum_{i}^{k} (x_{ni} - \mu)^2}{k}}, 1 \le i \le k \le D, \tag{4}$$

where μ represents the average electricity consumption over time. The dissipation coefficient of electricity consumption is recorded as dc, which indicates the degree of dispersion of the electricity consumption data, and its formula [33] is:

$$dc = \frac{std}{\mu}. \tag{5}$$

The difference between the mean values of electricity consumption in adjacent time intervals is avg_{ra} [33], which is:

$$avg_{di} = \left| \frac{\sum_{i=1}^{k} \overline{y}_{n(m+1)}}{k} - \frac{\sum_{i=1}^{k} \overline{y}_{n(m-i+1)}}{k} \right|. \tag{6}$$

The ratio between the mean values of electricity consumption in adjacent time intervals is avg_{ra}, which is:

$$avg_{ra} = \frac{\sum_{i=1}^{k} \overline{y}_{n(m+1)}}{k} \div \frac{\sum_{i=1}^{k} \overline{y}_{n(m-i+1)}}{k}. \tag{7}$$

The trend of electricity consumption rise and fall is obtained by comparing the actual value of electricity consumption x_{nt} at a certain time t with the predicted electricity consumption F_t at this time. Among them, the predicted value at a certain time is shifted item by item according to the time series through the simple moving average method, and its predicted value is the average value of the last fixed item number k. The F_t formula [33] is:

$$F_t = \frac{(x_{n(t-1)} + x_{n(t-2)} + \cdots + x_{n(t-k)})}{k}. \tag{8}$$

The rising and falling trend tr of a certain time t is:

$$tr = x_{nt} - F_t, \tag{9}$$

if $tr > 0$, it is an uptrend; if $tr < 0$, it is a downtrend.

Since the feature dimension of the above-mentioned extracted power consumption time series data is large, the features are redundant, and the feature matching is too complicated. Therefore, the extracted feature data needs to be dimensionally reduced. In this paper, principal component analysis (PCA) [34] is used to reduce the dimension of high-dimensional feature data, that is, a small number of new attributes are used to ensure that a large amount of original information is not lost. Suppose that the extracted power-time series data features are: $Y_{n \times f}$, where n is the number of samples and f is the feature dimension. The eigenvalues obtained by the PCA method are arranged from largest to smallest as follows: $\lambda = [\lambda_1, \lambda_2, \lambda_3, \ldots, \lambda_{f-1}, \lambda_f]$, and the matrix obtained by the eigenvectors corresponding to the previous l eigenvalues is $A_{f\times k}$. The new feature data obtained after calculating the dimensionality reduction of principal component analysis method are: $Y'_{n\times k} = Y_{n\times f} \times A_{f\times k}$, and the principal component contribution rate r is defined as the value criterion for l. The contribution rate r represents the proportion of the eigenvalues corresponding to the principal components in the data after dimensionality reduction, which reflects the reliability of the new features. In this paper, we choose $r \geq 95\%$ [34], that is:

$$r = \frac{\sum_{i=1}^{l} \lambda_i}{\sum_{i=1}^{f} \lambda_i} \geq 0.95, \tag{10}$$

where $l \leq f$.

3. Proposed Methods

This section details the paper's proposed design for the stacking integrated structure, followed by the electricity theft detection method based upon it, including the selection of the base-classifier model and the meta-classifier model, and the flow of the detection method.

3.1. Principles of Ensemble Learning

Ensemble learning accomplishes learning tasks through the construction and combination of multiple learners and can also be labeled a multi-classifier system. Figure 2 shows the usual architecture of ensemble learning. In essence, a set of single learners is first created, and these are then combined using a particular strategy. The single learners are usually derived from training data by a pre-designed learning algorithm. Ensemble learning, with its multiple combined learners, can often obtain significantly superior generalization performance and estimation accuracy than the single learner method.

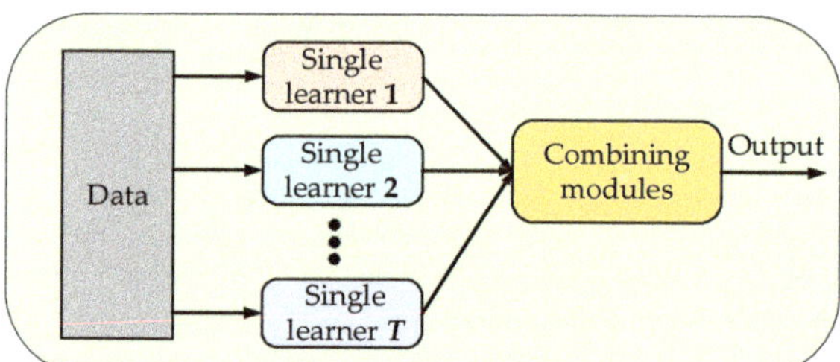

Figure 2. The usual structure of ensemble learning.

The most common ensemble methods include bagging, boosting, and stacking. Bagging trains homogeneous weak estimation models in parallel and averages the results from each one to achieve the final output. Boosting works similarly to bagging, but the weak models are given a variety of weights, so that the final output is given as weighted average values. In contrast, stacking creates its models through the use of different learning algorithms, which results in a unified methodology that can blend multiple estimation models into a single, unique metamodel. Stacking learning also has better generalization performance than other ensemble learning methods, as is corroborated in [35].

3.2. Stacking Integrated Structure

Stacking (sometimes called stacked generalization) was first introduced by David Wolpert in [35]. Its main purpose is to reduce generalization errors. According to Wolpert, stacked generalization can be understood as a "more complex version of cross-validation" that integrates models through a winner-takes-all approach.

The stacking integrated structure is composed of three parts. Firstly, the training data is evenly divided into k non-intersecting pieces as the data set for the classifiers' "leave- one-out" method training; secondly, the base classifiers are chosen from a number of classifiers, and their prediction results are obtained. Finally, the prediction results are used as the next stage feature input, a classifier is selected as a meta-classifier for training, and the prediction results are output. The integrated structure of stacking is depicted in Figure 3.

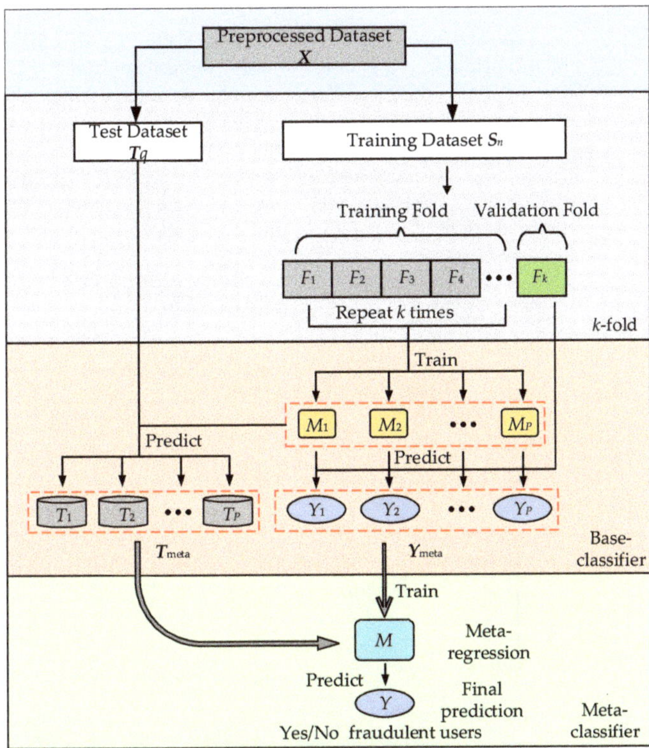

Figure 3. Structure of stacking integrated model.

For the first layer, the k-fold layer, the preprocessed dataset $\mathbf{X}$ is split between a training dataset and a test dataset, where the training dataset $S_n = \{(x_n, y_n), n = 1, 2, \ldots, N\}$ is divided into k-folds (i.e., $F_1, F_2, \ldots, F_k$), and the test dataset is $T_q = \{(x_q), q = 1, 2, \ldots, Q\}$. In S_n, x is the feature vector, and y is the classification attribute. The second layer, the base-classifier layer, contains P base models M_p (i.e., $M_1, M_2, \ldots, M_P$). For each base model $M_1, M_2, \ldots, M_P$, k training is performed separately, and $1/k$ samples are reserved for every training to be used as a test to make predictions. All prediction results are spliced, and $M_1, M_2, \ldots, M_P$ respectively get the meta training dataset $Y_{meta} = (Y_1, Y_2, \ldots, Y_P)$, while the result Y_P obtained by a single model is $Y_P = \{(y_{P1}, y_{P2}, \ldots, y_{Pk})\}$. Here, Y_{meta} actually refers to the meta-features of the training dataset [35].

$$\mathbf{Y}_{meta} = \begin{bmatrix} (y_{11}) & (y_{21}) & \cdots & (y_{P1}) \\ (y_{12}) & (y_{22}) & \cdots & (y_{P2}) \\ \vdots & \vdots & & \vdots \\ (y_{1k}) & (y_{2k}) & \cdots & (y_{Pk}) \end{bmatrix}, \tag{11}$$

Moreover, the base models $M_1, M_2, \ldots, M_P$ are trained k times each. The model obtained in each training is predicted on the test dataset, and the k prediction results of each model are averaged to obtain the meta test dataset $\boldsymbol{T}_{meta} = T_1, T_2, \ldots, T_P$, where $T_P = ((T_{P1}), (T_{P2}), \ldots, (T_{Pq}))$. As before, $\boldsymbol{T}_{meta}$ is the meta-features of the test dataset here [35].

$$\boldsymbol{T}_{meta} = \begin{bmatrix} T_{11}T_{21}\cdots T_{P1} \\ T_{12}T_{22}\cdots T_{P2} \\ \\ T_{1q}T_{2q}\cdots T_{Pq} \end{bmatrix}, \tag{12}$$

The last layer is the meta-classifier layer. A simple model is trained through the meta training dataset $\boldsymbol{Y}_{meta}$, and then the meta test dataset $\boldsymbol{T}_{meta}$ is predicted to get the final output. Since the base-classifier layer uses strong models to prevent over-fitting of the overall model, a simple one is generally chosen for the meta-classifier layer model. The linear regression model is a very common choice. In fact, a new simple model is used to train the super-features of the training dataset to train a model from meta-features to ground truth. Then, the meta-features of the test dataset are input into this model to obtain the final result. The pseudo-codes of the stacking integrated structure approach are given in Algorithm 1.

Algorithm 1: The *stacking* integrated structure

```
Input:  Trainning Dataset Sn = {(x1,y1), (x2,y2), ···, (xN,yN)}
        Base-classify Algorithm M1, M2, ···, MP
        Meta-classify Algorithm M
#Data processing
for p = 1, 2, 3, ···, P  do
    hp = Mp(Sn)
end for
Ymeta = ∅
for n = 1, 2, 3, ···, N  do
    for p = 1, 2, 3, ···, P  do
        Yp = Mp^(k)(xn)
    end for
    Ymeta = Ymeta ∪ (Y1, Y2, ···, YP)
end for
h' = M(Ymeta)
Output: Y = h'(Tmeta)
```

3.3. Flow of the Detection Method

The specific steps of the electricity theft detection process based on the stacking integrated structure experimental flow chart are shown in Figure 4. First, data are collected from smart meters, which form a historical electricity consumption dataset. The collected data are then preprocessed, including filling missing values and outlier removal (see Section 2.2 for details). Meanwhile, the pre-processed electricity consumption data is extracted for feature extraction in order to obtain better detection results. Finally, the data training and user prediction are carried out by establishing the stacking structure of the electricity theft detection model, including the selection and analysis of the base model, the selection and analysis of the metamodel, the selection of the super parameters in the classification model, and the optimization of the parameters of the metamodel through the algorithm to achieve the best detection effect.

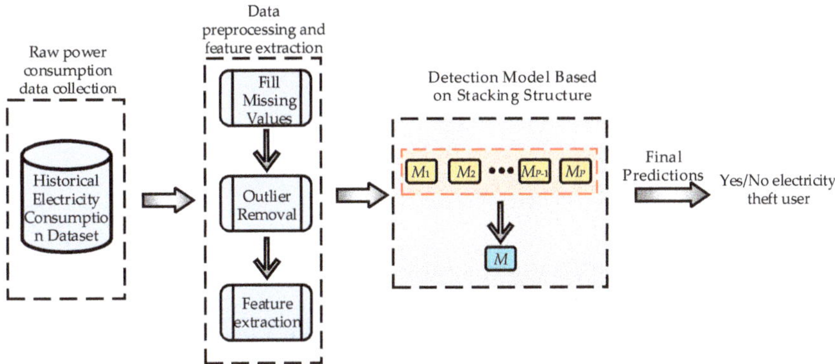

Figure 4. Structure of stacking integrated model experimental flow chart.

3.4. Selection of Stacking Structural Base Model and Meta Model

According to the previous introduction, this paper uses the combined weight method of subjective weight and objective weight based on GRA to determine the weight of a single performance index of the classifier and takes the final result of the weighted sum of each index as the base model evaluation criterion for selecting stacking structure.

Among them, common subjective methods of assigning weights include: expert survey method (Delphi method), analytic hierarchy method (AHP) [36], binomial coefficient method, chain comparison scoring method, least squares method, etc. Common methods of objectively assigning weights include: the entropy weight method (EWM) [37], principal component analysis method, factor analysis method, etc. According to the characteristics of the classifier evaluation index, the subjective assignment and weighting method selects a decision-making method with simple quantitative relationship and simple logic, namely analytic hierarchy process (AHP). The entropy weight method (EWM) is a more accurate method of objectively determining weights, which can supplement the subjective assignment and weighting method that is too subjective and insufficient, and the method can modify the determined weights, so its adaptability is stronger than other objective weighting and weighting methods.

Based on the GRA, the method of combining and assigning weights is based on the principle of the maximum gray correlation between subjective preference values and objective preference values and decision values, which has the characteristics of clear thinking, being concise and practical, and easy to implement on the computer.

First, the subjective weight value of the classifier performance index is determined by the AHP. In the field of electricity theft detection, the number of negative samples (i.e., samples of users who steal electricity) is much smaller than that of positive samples (i.e., normal user samples), so considering data redundancy, four relatively important evaluation criteria are selected as reference indicators, namely: Recall rate (Recall), MAP@100, F_1-score and AUC. In order to better introduce the above 4 indicators, we need to introduce a confusion matrix as shown in Table 3. The dataset provided in this paper is divided into normal users and thieving users and contains labels. The essence of theft detection is a binary classification problem.

Table 3. Confusion Matrix in the Detection of Electricity Theft.

Users	Detected as a Theft User	Detected as a Normal User
Theft users	*TP* (true positive)	*FN* (false negative)
Normal users	*FP* (false positive)	*TN* (true negative)

Recall rate (Recall) and F_1-score are defined using the confusion matrix in Table 3, corresponding to (13) and (14) [27]. F_1-score is the harmonic average of precision and Recall, which is able to comprehensively evaluate the performance of a classifier.

$$\text{Recall} = \frac{TP}{TP + FN}, \tag{13}$$

$$F_1\text{-score} = \frac{2TP}{2TP + FN + FP}, \tag{14}$$

A ROC (receiver operating characteristic) curve is used to express the relative relationship between FPR (FPR = $FP/(TN + FP)$) and TPR (TPR = $TP/(TP + FN)$) growth rates in the confusion matrix. In the ROC space, the closer coordinates are to the ROC curve on the upper left, the lower the FPR caused by the same detection rate, and the better the detection performance. AUC (area under ROC curve) is the sum of the areas under the ROC curve. For the purpose of comparing each classifier's performance, the larger the AUC value, the better, and when AUC = 1, the classifier is ideal. The calculation formula of AUC is as follows [27]:

$$\text{AUC} = \frac{\sum_{i \in \text{positive}} \text{Rank}_i - \frac{H(1+H)}{2}}{H \times F}, \tag{15}$$

where Rank_i signifies the ranking value of sample i, H signifies the number of positive samples, and F signifies the number of negative samples.

Mean average precision (MAP) is used to evaluate the performance of model detection. MAP@F is defined as the average accuracy of the detection model correctly identified as thieving users among the top F users with the highest suspicion. MAP@F is as follows [27]:

$$\text{MAP@F} = \frac{\sum_{i=1}^{r} P@k_i}{r}, \tag{16}$$

where r represents the number of users who steal electricity among the top F users with the highest suspicion; $P@k_i$ is defined as [27]:

$$P@ki = \frac{Y_{ki}}{ki}, \tag{17}$$

where Y_{ki} represents the number of users who are correctly identified electricity thieves among the first k users with the highest suspicion, and k_i ($i = 1, 2, 3, \ldots, r$) represents the position of k. In this paper, we use MAP@100 as evaluation metrics.

The higher the Recall, the lower the number of users who steal electricity and are misidentified as normal, so this metric has a greater impact on the model. MAP@100 is in the first 100 users with the highest suspicion, the detection model is correctly identified as the average accuracy of the electricity theft user, if the prediction result of the classifier is all judged to be the electricity theft user, then the Recall is very high and the accuracy rate is very low, this result is not conducive to distinguishing between normal users and electricity theft users, and MAP@100 is an important supplement to Recall, so its importance is higher than Recall. F_1-score is the harmonic mean of Recall and accuracy, and the higher the value, the more credible the classification result, so its importance is higher than MAP@100. AUC can be obtained by summing the areas of the parts under the ROC curve, the larger the AUC value, the better, and the ideal classifier is obtained when AUC = 1. Therefore, AUC is the most important in the pursuit of the accuracy of electricity theft detection. The weight values of Recall, MAP@100, F_1-score, and AUC obtained according to the AHP are shown in Table 4.

Table 4. The AHP method determines the classifier performance metric weight value.

Metrics	Recall	F_1-Score	MAP@100	AUC
Recall	1	1/5	1/4	1/6
F_1-score	5	1	2	1/2
MAP@100	4	1/2	1	1/3
AUC	5	2	3	1

Next, the objective weight value of the classifier performance index is determined by the EWM. The EWM mainly determines the weight according to the amount of information transmitted to the decision-maker by each evaluation index, and is a mathematical method for calculating comprehensive indicators. Assuming that the base model is m classifiers or a combination of classifiers, the evaluation index reflecting its model is n. Let $X = \{x_1, x_2, \ldots, x_m\}$ represent the set of schemes for multi-attribute decision problems, $G = \{G_1, G_2, \ldots, G_n\}$ represents its corresponding set of properties, and $w = (w_1, w_2, \ldots, w_n)^T$ represents its corresponding property weight vector. Remember the decision matrix $R = (r_{i,j})_{m\times n}$, where $r_{i,j}$ is the decision value of the ith classifier on the j indicator. Calculate the information entropy of the j indicator H_j [36]:

$$H_j = -\frac{1}{ln(m)}\sum_{i=1}^{m} p_{i,j} ln(p_{i,j}), \tag{18}$$

where $p_{i,j} = \frac{r_{i,j}}{\sum_{i=1}^{m} r_{i,j}}$ represents the proportion of each metric for a classifier to the total statistical value of that metric, $0 < H_j < 1$. According to the entropy value H_j of each indicator, the entropy weight w_j of the corresponding indicator can be determined [36]:

$$w_j = \frac{1 - H_j}{\sum(1 - H_j)}. \tag{19}$$

It can be seen from the entropy weight w_j that when the value of each classifier differs on the indicator, the smaller the information entropy and the greater its entropy weight, which means that the indicator can provide more useful information to the decision maker.

Finally, the subjective weight values determined by the above hierarchical analysis and the objective weight values determined by the entropy method are combined by the combined empowerment method based on the grey correlation degree analysis method. The specific calculation steps are:

In the first step, according to the decision matrix R, the relationship between the comprehensive attribute value Z_i and the attribute weight of the scheme x_i is [32]:

$$Z_i = \sum_{j=1}^{n} r_{i,j} w_j, i \in M. \tag{20}$$

In the second step, the weight vector w' of the attribute is obtained by using the AHP, and the weight vector w'' of the attribute is obtained by using the EWM. Formula (20) is used to obtain the subjective preference value Z' and the objective preference value Z'' of each scheme. Before calculating the grey correlation coefficient, the parent and sub-indicators need to be determined. The parent indicator is $X_0 = (x_{1,0}, x_{2,0}, \ldots, x_{m,0})^T$. Other factor indicators, i.e., sub-indicators, are denoted as $X_j = (x_{1,j}, x_{2,j}, \ldots, x_{m,j})^T$, where $j = 1, 2, \ldots, n$. Calculate the gray correlation coefficient $\delta_{i,j}$ for X_0 and X_j [32]:

$$\delta_{i,j} = \frac{\underset{1\le j\le n}{min}\underset{1\le i\le m}{min}\left|\Delta_{i,j}\right| + \rho\underset{1\le j\le n}{max}\underset{1\le i\le m}{max}\left|\Delta_{i,j}\right|}{\left|\Delta_{i,j}\right| + \rho\underset{1\le j\le n}{max}\underset{1\le i\le m}{max}\left|\Delta_{i,j}\right|}, \tag{21}$$

where $\Delta i,j = x_{i,0} - x_{i,j}$, ρ represents the resolution coefficient, the value of which is $\rho \in [0, 1]$, which is generally $\rho = 0.5$. According to Equation (20), the gray correlation coefficient $\delta i,j$ between the subjective preference value Z' and the objective preference value Z'' and the decision value $r_{i,j}$ (where the former is $\delta'_{i,j}$, and the latter is $\delta''_{i,j}$). The grey correlation coefficient $\delta_{i,j}$ reflects the similarity between the objective preference and subjective preference of the decision maker for indicator j and the decision value, and the larger the value of $\delta_{i,j}$ indicates that the subjective preference and objective preference of the decision maker for indicator j are more similar to the decision value.

In the final step, since the various schemes are fairly competitive, that is, no preference for any of them, the following objective optimization model can be established [37]:

$$\begin{cases} max\delta_{i,j} = \sum\limits_{j=1}^{n}\sum\limits_{i=1}^{m}(\delta\prime_{i,j} + \delta''_{i,j})W_j \\ s.t.W_j \in w, W_j > 0, \sum\limits_{j=1}^{n} W_j = 1 \end{cases}. \tag{22}$$

According to the above optimization model, the combined weight vector W_j can be solved.

For the final result of the weight vector W_j weighted sum of m classifiers or classifiers obtained by the analysis of GRA, η used as the base model evaluation criterion for selecting stacking structure, in which a classifier or combination of classifiers with relatively large comprehensive evaluation index values is selected as the base model. In the process of feature extraction, due to the use of complex nonlinear transformations, complex classifiers are not required at the metamodel layer, but a simpler model is selected to prevent overfitting of the overall model. The model selection principle is a classifier that is simple and has good classification prediction results [35].

4. Evaluations

In order to authenticate the effectiveness and accuracy of the algorithm given in this paper, it should be noted that the experimental hardware is a 64-bit, 6-core Intel Core i7-8750H CPU@2.20 GHz, and the deep learning framework uses TensorFlow and Keras. The programming accomplished using PyCharm 2020 (The software version number is: Pycharm2020.3.2, developed by JetBrains, headquartered in Prague, Czech Republic). The experimental data used in this paper are based on a dataset from a province of the State Grid Corporation of China (refer to Section 2.1 of this paper).

4.1. Construction of Base Model Layer in Stacking Structure

According to the experimental flow of the electricity theft detection model based on stacking structure in Figure 4, the data preprocessing, including missing value complement and outlier value repair, has been described in detail in Section 2 of the article, and the principle of feature extraction (i.e., load sequence feature extraction) for electricity consumption data has been described in detail in Section 2.3 of the article, where the load sequence feature extraction is performed on the SGCC dataset to obtain time series features $[D_1, \ldots, D_{49}]$. The newly dimensionality-reducing features of the extracted high-dimensional time series $[D_1, \ldots, D_{49}]$ were then treated by the PCA method described above to obtain the new dimensionality-reducing feature values from largest to smallest: $\lambda = [\lambda_1, \lambda_2, \lambda_3, \ldots, \lambda_{48}, \lambda_{49}]$. Calculate the value of l when the principal component contribution rate $r \ge 95\%$ is calculated by Formula (10), and $l = 6$ is obtained after calculation, that is, the first six principal component eigenvalues are selected as the new feature set Y.

The selection of the base model layer and the metamodel layer in Figure 3 is the most important part of building the stacking structure, and the principle of the selection of the base model layer and the metamodel layer has been described in detail in Section 3.3, where the base model layer is more complex than the metamodel because of the large number of classifiers in this layer. The base model layer determines the weight values of a single performance index of the classifier by using a combined weight method of subjective weights and objective weights based on GRA, of which the subjective weight method obtains the weights of Recall, MAP@100, F_1-score and AUC through the AHP as shown in Table 4, and the weights of the four indicators are further calculated to be: $w' = (0.0598, 0.2933, 0.1786, 0.4683)$.

In order to obtain the objective weight w'' obtained by the EWM, the decision matrix $R = (r_{i,j})_{m\times n}$ of each classifier or a combination of classifiers (that is, each scheme) is first required, that is, the different classifications in the stacking structure are selected. The combined base model has four performance indicators: Recall, MAP@100, F_1-score, and AUC under the new feature set Y after preprocessing, feature extraction and dimensionality reduction of the SGCC dataset, at this time, the meta-model of the stacking structure chooses a relatively simple linear regression (LR) model [38]. According to the classifier, selection of the base model layer, as in Section 3.3, should be strong and numerous, so the performance index values of eight existing classifiers commonly used for electricity theft detection under the new feature set Y are compared, and the eight classifiers are: random forest (RF) [39], eXtreme gradient boosting (XGBoost) [25], light gradient boosting machine (LightGBM) [40], support vector machine (SVM) [22], CART decision tree (DT) [23], deep forest (DF) [41], long short-term memory (LSTM) [28], and K-nearest neighbor (KNN) [42].

The hyperparameters of the above eight classifier algorithms are set to: In the RF model, the number of decision trees and the maximum depth of the tree are set to 101 and 15, respectively. The XGBoost model sets the learning rate to 0.5, the random sampling ratio to 0.08, and the maximum depth and optimal number of iterations to 3 and 10, respectively. The LightGBM model sets the number of leaf nodes to 10, the learning rate to 0.05, the feature selection scale and sample sampling ratio of the tree to 0.8, and the number of iterations required to perform bagging is 5. The SVM model sets the kernel function as a radial basis function, and the penalty coefficient $C = 15$. The DT model sets the confidence parameter $\theta = 0.25$, the minimum number of instances on the leaf node $\rho = 2$. The number of decision trees required for the DF model to set up multi-granular scanning is $K = 30$, and the slicing window size is 15. The LSTM model sets the number of neurons to 32, the number of hidden layers to 2, the learning rate to 0.1, and the number of trees to 300. The KNN model sets the initial K value to 3.

The new feature set Y data samples are divided, and 50% of the data is randomly selected as the training sample (corresponding to 50% of the data as the test sample), and Table 5 is the experimental results of the above eight classifiers, that is, the decision matrix R. Therefore, the objective weight method obtains call, MAP@100, F_1-score, and AUC through the EWM, and the four performance index weights are: $w'' = (0.25899, 0.24321, 0.24851, 0.24929)$.

The combined weight vectors of each index of the combined weighting method can be obtained in three steps based on GRA: $W_j = [0.0598, 0.2432, 0.2287, 0.4683]$. According to the combined weight vector W_j, the comprehensive evaluation index values of the above eight classifiers are calculated: $\eta_1 = [0.8273, 08107, 0.7991, 0.7318, 0.6863, 0.7962, 0.8110, 0.6848]^T$, from which the comprehensive evaluation index of the above 8 classifiers is sorted as: RF > LSTM > XG > LG > DF > SVM > DT > KNN. The classifiers of the base model layer are combined according to the above eight classifiers, and the classifier combinations are combined from 2 to 8, where the number of combination types is: $C_8^2+C_8^3+C_8^4+C_8^5+C_8^6+C_8^7+C_8^8 = 247$, due to the many combinations, as shown in Table 6.

Table 5. The experimental results of 8 classifiers under feature set Y.

Classifier	Metrics			
	Recall	F_1-Score	MAP@100	AUC
RF	0.87831	0.85061	0.86121	0.79187
XG	0.87483	0.84731	0.81756	0.78112
LG	0.86815	0.84441	0.81261	0.76108
SVM	0.86743	0.82417	0.76136	0.64407
DT	0.85911	0.79401	0.63899	0.63625
DF	0.72667	0.84617	0.82528	0.76551
LSTM	0.85928	0.83304	0.85928	0.76909
KNN	0.86001	0.79529	0.61371	0.64538

Table 6. The experimental results of each classifier combination under feature set Y.

Number of Classifiers	The Combination of Classifiers	Metrics			
		Recall	F_1-Score	MAP@100	AUC
2	(DF + LSTM) [i]	0.89598	0.88095	0.92766	0.84267
	(XG + LSTM) [ii]	0.90143	0.88937	0.94245	0.84268
	(LG + LSTM) [iii]	0.90341	0.89259	0.95528	0.85764
3	(DF + LSTM + KNN) [iv]	0.98431	0.91358	0.99378	0.94881
	(XG + LSTM + KNN) [v]	0.98642	0.98637	0.99872	0.95149
	(LG + LSTM + KNN) [vi]	0.98712	0.99872	0.99969	0.97401
4	(DF + LSTM + KNN + SVM) [vii]	0.98599	0.91531	0.99667	0.95691
	(XG + LSTM + KNN + SVM) [viii]	0.98945	0.98431	0.99378	0.95841
	(LG + LSTM + KNN + SVM) [ix]	0.98761	0.99898	0.99979	0.97659
5	(DF + LSTM + KNN + SVM + XG) [x]	0.97185	0.90857	0.98011	0.93027
6	(DF + LSTM + KNN + SVM + XG + LG) [xi]	0.96493	0.91571	0.97401	0.92857
7	(DF + LSTM + KNN + SVM + XG + LG + RF) [xii]	0.95944	0.91385	0.96815	0.92779
8	(DF + LSTM + KNN + SVM + XG + LG + RF + DT) [xiii]	0.94521	0.91706	0.96529	0.92262

The experimental results only list some valuable classifier combinations (each quantity combination classifier selects relatively good displays according to the performance index values) and its corresponding Recall, MAP@100, F_1-score, and AUC of the four performance index values. At this point, the meta-model of the stacking structure selects a linear regression model, and the k-fold setting $k = 5$.

The comprehensive evaluation index values of stacking structure integration learning method of each of the above classifier combinations were calculated by the combined weight vector W_j based on gray correlation degree analysis, and the results were: η_2 = [0.8746, 0.8804, 0.8912, 0.9526, 0.9729, 0.9867, 0.9576, 0.9746, 0.9879, 0.9389, 0.9380, 0.9355, 0.9324]^T. The η_2 corresponds to the comprehensive evaluation index values of each of the above classifier combinations, from which the comprehensive evaluation indexes of the above 13 classifier combinations can be sorted as follows: ix > vi > viii > v > vii > iv > x > xi > xii > xiii > iii > ii > i, that is, the top three combinations of the comprehensive evaluation index values of the 13 classifier combinations are: LG + LSTM + KNN + SVM, LG + LSTM + KNN and XG + LSTM + KNN + SVM, the corresponding comprehensive evaluation index values are 0.9879, 0.9867 and 0.9746, respectively. So, the stacking structure integration learning method of the three classifiers combination base model layers has a better effect on the detection and classification of electricity theft behavior.

Through the above method, three classifier combinations with relatively good comprehensive evaluation index values were selected, but the comprehensive evaluation index values were relatively close (the difference was about 0.001). In order to select the optimal classifier combination, the training time of the model is also an important reference index for the real-time detection of electricity theft, so the training time of the stacking structure integration learning model under different classifier combinations is considered (at this time, the metamodel still uses a linear regression model). As shown in Figure 5, given the training time of the stacking structure integration learning model under different classifier combinations, it can be clearly concluded that when the base model layer adopts the LG + LSTM + KNN combination, the model training time of the stacking structure is the least, only 13.078 s. The longest model training time is the XG + LSTM + KNN + SVM combination, and the training time is 17.154 s.

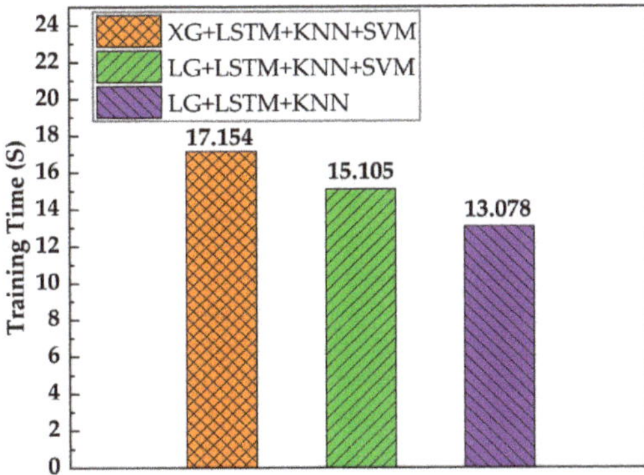

Figure 5. Training time of stacking structure integration learning model under different classifier combinations.

Taking into account the accuracy of the model and the training time of the model, the base model layer of the stacking structure ensemble learning model selects LG + LSTM + KNN. The comprehensive evaluation index value of stacking structure ensemble learning model detection based on this base model layer is only 0.0012 different from the comprehensive evaluation index value of stacking structure ensemble learning model detection based on the combination of LG + LSTM + KNN + SVM. The training time difference is 2.027 s. Therefore, considering the above factors, the combination of LG + LSTM + KNN is selected as the base model of the stacking structure ensemble learning model.

The above experiments set $k = 5$ in the k-fold layer, and different k values will greatly affect the detection effect of the stacking structure. According to the above experiments, the combination of LG + LSTM + KNN is selected as the base model of the stacking structure ensemble learning model, and the linear regression model is selected for the meta-model layer, and the k values are set to 2, 3, 4, 5, 10, 15, and 20 pairs of models respectively. After training, Figure 6 shows the experimental results under different k values, in which the experimental results are the four performance index values of Recall, MAP@100, F_1-score, and AUC. As can be seen from Figure 6, as the value of k increases, the values of the four performance indicators also increase. When the value of k is 5, each indicator value reaches the maximum value. On the other hand, the experimental results with k-fold cross-training are better than those without k-fold cross-training, so k-fold cross-training significantly improves the detection performance of the model. Therefore, when the combination of LG + LSTM + KNN constitutes a stacking structure, five-fold cross-training is selected, that is, $k = 5$ is set as the best parameter in the k-fold layer.

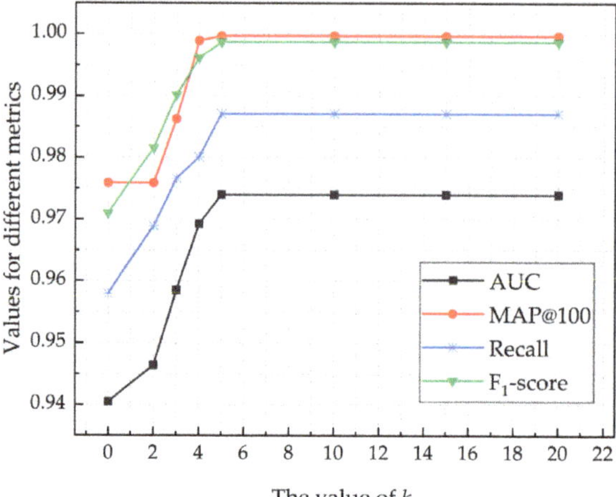

Figure 6. Experimental results of stacking structure with different k values.

The stacking structure integration learning method integrates a variety of detection algorithms, which can make full use of each algorithm to observe data from different data spaces and structures. Therefore, the classifier of the base model layer should try to choose an algorithm with excellent performance and should also add different types or properties of classification algorithms as much as possible. In order to further verify and analyze the optimal base model combination selected, each base learner separately compares the classification prediction of the new feature set Y, and the Pearson correlation coefficient matrix is used to analyze the correlation of the classification prediction index values (AUC), and its calculation formula is as follows [33]:

$$r = \frac{\sum_{i=1}^{n}(x_i - \overline{x})(y_i - \overline{y})}{\sqrt{\sum_{i=1}^{n}(x_i - \overline{x})^2 (y_i - \overline{y})^2}}, \tag{23}$$

where $\overline{x}, \overline{y}$ is the sample mean. The larger value of $|r|$, the more correlated it is. Figure 7 shows the correlation coefficient matrix between each classifier.

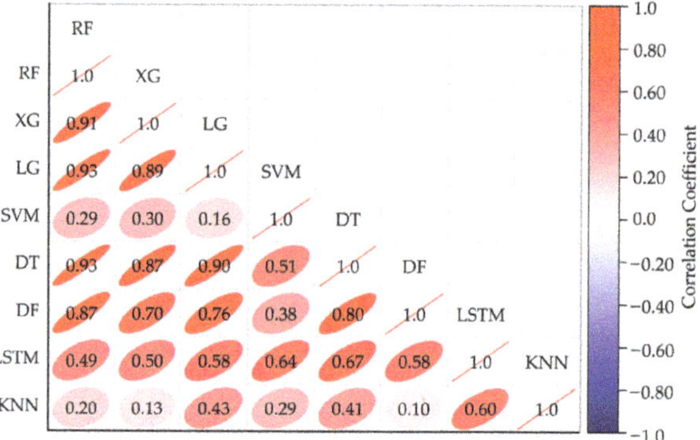

Figure 7. Matrix of correlation coefficients for the value of the classification prediction indicator between classifiers.

It can be seen from Figure 7 that the correlation degree of the value of the classification prediction index of each algorithm is generally high, which is due to the strong learning ability of each algorithm, and the inherent laws learned in the data during training are similar to the data observation angles. Among them, the classification prediction index values of RF, XG, LG, DF, and DT algorithms have the highest correlation, which is due to the fact that although the principles of the five types of algorithms are slightly different, they still belong to the integrated algorithms of the tree, and there are strong similarities in their data observation methods. The training mechanisms of LSTM, SVM, and KNN have a large gap, so the correlation of classification prediction index values is also low. Therefore, the effectiveness of the base model layer in choosing LG + LSTM + KNN algorithm combination as the base model in stacking integration learning is further verified.

4.2. Construction of Meta Model Layer in Stacking Structure

As described in Section 3.2, the meta-model layer usually chooses a relatively simple model to prevent the overfitting problem of the collation model, so this section selects several relatively simple models at the meta-classifier layer to compare the experimental results of the stacking structural integration learning method, including the SVM, DT, KNN, and LR. The ROC curves of the experimental results of the stacking structure under the above four different meta-models are shown in Figure 8.

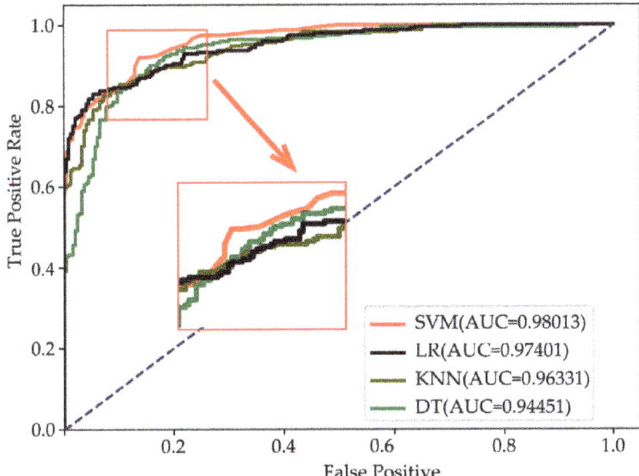

Figure 8. The ROC curve of stacking structures under different meta-models.

It can be clearly seen from Figure 8 that when SVM is selected for the meta-model layer, the overall detection effect of Stacking ensemble learning is the best, and its AUC value is 0.98013. When the meta-model layer adopts decision tree, the sorting and detection effect of the stacking ensemble learning is slightly worse than the other three. Therefore, considering the detection effect, this paper adopts SVM as the model of the stacking integrated learning meta-model layer.

Since the recognition accuracy of the SVM algorithm is limited to a large extent by the selection of parameters, and the parameter optimization algorithm generally has problems, such as slow convergence speed and a tendency to fall into local extremums, the particle swarm optimization (PSO) algorithm [43] with strong optimization ability, fast convergence speed, and short calculation time is selected in this experiment to optimize the penalty coefficient C and kernel function (i.e., radial basis function) parameter σ values in the stacking integrated learning model metaclassifier SVM hyperparameter. Figure 9 shows the particle swarm optimization metaclassifier SVM hyperparameter flowchart, which is implemented as follows:

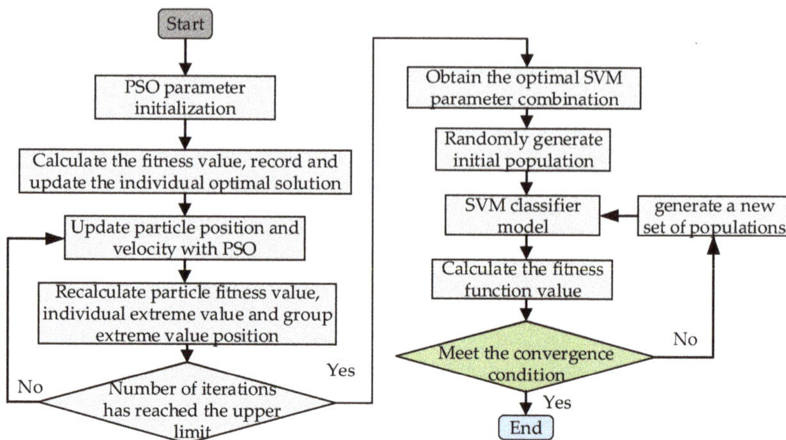

Figure 9. Flowchart of PSO to optimize meta-classifier (SVM) hyperparameters.

First of all, the initialization stage of the PSO parameter sets the step size and upper and lower boundaries of the search parameters, and the local optimal solution of the particles, the global optimal solution of the particle swarm, and its corresponding position are obtained by calculating the fitness function, and the fitness function adopts the cross-validation scores (CVS) method, which is calculated as follows [43]:

$$\text{CVS} = \frac{1}{k}\sum_{i=1}^{k}\frac{y_i}{y}, \tag{24}$$

where k is the number of cross-validations, y represents the number of training samples, y_i is the number of training samples that are correctly divided, and the higher the CVS value, the higher the accuracy of the model.

Second, the velocity and position of the individual particle swarm are iteratively updated according to the local optimal and global optimal solutions, and the cycle is reached until the maximum number of iterations is reached.

Finally, the parameters corresponding to the global optimal particle swarm individuals obtained above are trained as the initial parameters of the SVM, and the fitness value of each particle is calculated by the k-fold cross-validation value method again. If the adaptability of the new position is better than that of the local optimal particle, the local optimal particle is replaced with the new particle. If the optimal particle in the population is superior to the global optimal particle, the global optimal particle is replaced by the best particle in the population. The global optimal parameter C and σ values are returned after the above is completed.

The above particle swarm algorithm optimizes the stacking ensemble learning model meta-classifier SVM hyperparameter, and the basic parameters of PSO setting are: acceleration factor c_1 and c_2 are both 2, inertia factor ω = 1, the number of particle swarms is 20, and the maximum number of iterations is 50. Figure 10 shows an evolutionary iteration plot that represents the resulting change in fitness values over different evolutionary algebras. As can be seen from Figure 10, PSO optimizes the SVM process, the fitness value remains unchanged after 26 iterations, and the final optimal fitness value is 0.976013, at which time the optimal parameter combination of the trained SVM is C = 21.375 and σ = 1.43.

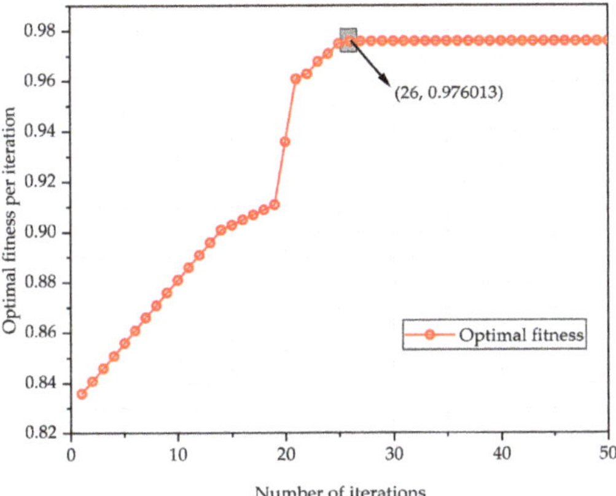

Figure 10. Diagram of evolutionary iterations.

When the PSO optimization SVM obtains the best adaptability value, the AUC value is compared with the different effects before and after the optimization of the SVM parameters, as shown in Figure 11, which is the ROC curve of the stacking integration learning model before and after optimization.

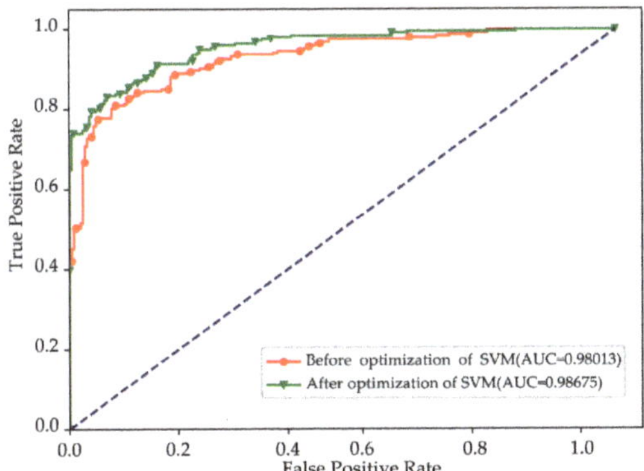

Figure 11. ROC curve before and after optimization.

It can be clearly seen from the ROC curve that the AUC value before optimization is 0.98013, while that after optimization is 0. 98675, and the AUC value is increased by about 0.007, because the detection effect of the stacking integrated learning model is relatively satisfactory, and the room for improvement is effective. So, SVM can relatively effectively improve the overall performance of the algorithm.

4.3. Comparison with Existing Methods

In order to verify the effectiveness of the detection method of stealing behavior based on the stacking ensemble learning model proposed in this paper, the experimental results are compared by using CNN [44], wide and deep CNN [27], hybrid deep neural networks (HDNNs) [45], CNN-RF [39] and the methods adopted in this paper. The dataset used in the above method is described in Section 3.1. Figure 12 shows the ROC curve of the above five methods, and the experimental results of the above five methods are shown in Table 7.

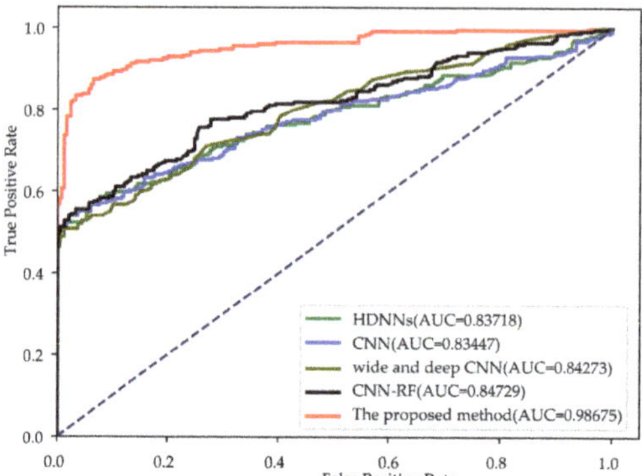

Figure 12. The ROC curves of the method proposed in this paper and the other four methods.

Table 7. The methods proposed in this article compare the results with other methods.

Methods	Metrics			
	Recall	**F_1-Score**	**MAP@100**	**AUC**
CNN	0.82613	0.75625	0.86015	0.83447
wide and deep CNN	0.85862	0.86331	0.87329	0.84273
Hybrid Deep Neural Networks(HDNNs)	0.84228	0.86085	0.86265	0.83718
CNN-RF	0.87637	0.89628	0.91358	0.84729
The proposed method	0.98948	0.99913	0.99975	0.98675

We can see from Table 7 that the evaluation indicators of the method proposed in this paper under the actual power grid data are better than the other four existing detection methods, of which the AUC value is 0.98675, which is much higher than the other four methods.

In addition, for Recall and F_1-score, the method in this paper is one order of magnitude higher than other methods. For example, the Recall of this method reaches 0.98948, while the highest Recall value of the other four methods is CNN-RF, which is 0.87637. In addition, we found that the other four methods are all deep learning methods, three of which are variants of CNNs, that is, optimization on CNNs. Compared with the automatic extraction process of CNN, the purpose of manual feature extraction and selection of the proposed method is more clear and more efficient. Moreover, the stacking structure is a combination of multiple strong models that can learn from different angles of the data, and the learning ability of this method is stronger.

In summary, the evaluation indicators of CNN and its optimization methods have been improved to a certain extent, but they still cannot reach a very high level. It is worth noting that the method proposed in this paper can break through the bottleneck where other methods cannot improve after the accuracy reaches a certain level and achieve the purpose of improving the accuracy rate.

5. Conclusions

In this paper, we propose a multi-model fusion ensemble learning algorithm based on the stacking structure to detect electricity theft behaviors. The feature of this paper based on the stacking structure detection algorithm is that the PCA method is used to reduce the dimensionality of the user time series statistical feature indicators in the extracted dataset. That is, only the new properties of the first six principal component eigenvalues are used to ensure that a large amount of original information is not lost. The subjective weight values determined by the AHP method and the objective weight values determined by the EWM are combined and weighted by GRA method. The classifier combination of LG + LSTM + KNN with a relatively high comprehensive evaluation value (0.9867) and a relatively low model training time (13.078 s) is selected as the base model layer of the stacking structure by comprehensive evaluation index values through a large number of experiments.

In the meta-model layer, several relatively simple models are selected for comparative experiments. The SVM model with relatively good overall structure experimental results (the AUC value is 0.98013) of stacking is selected as the meta-model. The PSO algorithm is used to optimize the hyperparameters of the SVM model and improve the AUC value of the model from 0.98013 to 0.98675. By comparing the stacking structural model with the existing methods under the SGCC dataset, the effectiveness of the proposed methods is further verified. For example, the AUC value of the method proposed in this paper is 0.98675, which is an order of magnitude higher than the CNN-RF method with the highest AUC value of 0.84729 among other methods. Therefore, the stacking structure integrated learning method can effectively realize the accurate detection and identification of electricity theft behavior.

Author Contributions: Conceptualization, R.X.; Data curation, R.X.; Funding acquisition, Y.G.; Methodology, R.X.; Project administration, Y.G.; Resources, R.X.; Software, R.X.; Supervision, D.G. and J.W.; Validation, R.X.; Visualization, R.X.; Writing—original draft, R.X.; Writing—review & editing, Y.G., Y.Z., D.G. and J.W. All authors have read and agreed to the published version of the manuscript.

Funding: This work is supported by National Natural Science Foundation of China (NO. 51777061) and 2020 Science and Technology Project of China Southern Power Grid Guangxi Electric Power Company (NO. GXKJXM20200020).

Conflicts of Interest: The authors declare no conflict of interest.

Nomenclature

AEA	Auto-encoder with attention
AHP	Analytic hierarchy process
AUC	Area under ROC curve
CNN	Convolutional Neural Network
CVS	Cross validation scores
DF	Deep forest
DT	Decision tree
EWM	Entropy weight method
F_1-score	The harmonic average of *precision* and *Recall*, which is able to comprehensively evaluate the performance of a classifier
FN	False negative
FP	False positive
FPR	False positive rate
GRA	Grey relation analysis
GRUs	Gated recurrent units
KNN	K-Nearest Neighbor
LG	Light gradient boosting machine, LightGBM
LR	Linear regression
LSTM	Long Short-Term Memory
MAP	Mean average precision
PCA	Principal component analysis
PSO	Particle swarm optimization
RF	Random forest
ROC	Receiver operating characteristic
$Rank_i$	The ranking value of sample i

SGCC	State Grid Corporation of China
SSAE	Semi-Supervised AutoEncoder
SVM	Support vector machine
TN	True negative
TP	True positive
TPR	True positive rate
XGBoost	eXtreme gradient boosting

References

1. Xia, X.; Xiao, Y.; Liang, W.; Cui, J. Detection Methods in Smart Meters for Electricity Thefts: A Survey. *Proc. IEEE* **2022**, *110*, 273–319. [CrossRef]
2. Saeed, M.S.; Mustafa, M.W.; Hamadneh, N.N.; Alshammari, N.A.; Sheikh, U.U.; Jumani, T.A.; Khalid, S.B.A.; Khan, I. Detection of Non-Technical Losses in Power Utilities—A Comprehensive Systematic Review. *Energies* **2020**, *13*, 4727. [CrossRef]
3. Xia, X.; Xiao, Y.; Liang, W.; Zheng, M. GTHI: A Heuristic Algorithm to Detect Malicious Users in Smart Grids. *IEEE Trans. Netw. Sci. Eng.* **2020**, *7*, 805–816. [CrossRef]
4. Feng, X.; Hui, H.; Liang, Z.; Guo, W.; Que, H.; Feng, H.; Yao, Y.; Ye, C.; Ding, Y. A Novel Electricity Theft Detection Scheme Based on Text Convolutional Neural Networks. *Energies* **2020**, *13*, 5758. [CrossRef]
5. Park, C.H.; Kim, T. Energy Theft Detection in Advanced Metering Infrastructure Based on Anomaly Pattern Detection. *Energies* **2020**, *13*, 3832. [CrossRef]
6. Xiong, D.; Chen, Y.; Chen, X.; Liu, X.; Yang, M. Design and Application of Intelligent Electricity Monitoring Device. In Proceedings of the 2018 International Conference on Power System Technology (POWERCON), Guangzhou, China, 6–8 November 2018; pp. 3312–3317.
7. Pamir; Javaid, N.; Javaid, S.; Asif, M.; Javed, M.U.; Yahaya, A.S.; Aslam, S. Synthetic Theft Attacks and Long Short Term Memory-Based Preprocessing for Electricity Theft Detection Using Gated Recurrent Unit. *Energies* **2022**, *15*, 2778. [CrossRef]
8. Raggi, L.M.; Trindade, F.C.; Cunha, V.C.; Freitas, W. Non-Technical Loss Identification by Using Data Analytics and Customer Smart Meters. *IEEE Trans. Power Del.* **2020**, *35*, 2700–2710. [CrossRef]
9. Leite, J.B.; Mantovani, J.R.S. Detecting and Locating Non-Technical Losses in Modern Distribution Networks. *IEEE Trans. Smart Grid* **2018**, *9*, 1023–1032. [CrossRef]
10. Zanetti, M.; Jamhour, E.; Pellenz, M.; Penna, M.; Zambenedetti, V.; Chueiri, I. A Tunable Fraud Detection System for Advanced Metering Infrastructure Using Short-Lived Patterns. *IEEE Trans. Smart Grid* **2019**, *10*, 830–840. [CrossRef]
11. Guerrero, J.I.; Monedero, I.; Biscarri, F.; Biscarri, J.; Millan, R.; Leon, C. Non-Technical Losses Reduction by Improving the Inspections Accuracy in a Power Utility. *IEEE Trans. Power Syst.* **2018**, *33*, 1209–1218. [CrossRef]
12. Wei, L.; Sundararajan, A.; Sarwat, A.I.; Biswas, S.E.; Ibrahim, E. A distributed intelligent framework for electricity theft detection using benford's law and stackelberg game. In Proceedings of the 2017 Resilience Week (RWS), Wilmington, DE, USA, 18–22 September 2017; pp. 5–11.
13. t Chen, Q.; Zheng, K.; Kang, C.; Huang, F. Detection Methods of Abnormal Electricity Consumption Behaviors: Review and Prospect. *Autom. Electr. Power Syst.* **2018**, *42*, 189–199.
14. Amin, S.; Schwartz, G.A.; Cardenas, A.A.; Sastry, S.S. Game-Theoretic Models of Electricity Theft Detection in Smart Utility Networks: Providing New Capabilities with Advanced Metering Infrastructure. *IEEE Control Syst. Mag.* **2015**, *35*, 66–81.
15. Gao, Y.; Foggo, B.; Yu, N. A Physically Inspired Data-Driven Model for Electricity Theft Detection With Smart Meter Data. *IEEE Trans. Ind. Informat.* **2019**, *15*, 5076–5088. [CrossRef]
16. Zheng, K.; Chen, Q.; Wang, Y.; Kang, C.; Xia, Q. A Novel Combined Data-Driven Approach for Electricity Theft Detection. *IEEE Trans. Ind. Informat.* **2019**, *15*, 1809–1819. [CrossRef]
17. Takiddin, A.; Ismail, M.; Zafar, U.; Serpedin, E. Robust Electricity Theft Detection Against Data Poisoning Attacks in Smart Grids. *IEEE Trans. Smart Grid* **2021**, *12*, 2675–2684. [CrossRef]
18. Zhang, Q.; Zhang, M.; Chen, T.; Fan, J.; Yang, Z.; Li, G. Electricity Theft Detection Using Generative Models. In Proceedings of the 2018 IEEE 30th International Conference on Tools with Artificial Intelligence (ICTAI), Volos, Greece, 5–7 November 2018; pp. 270–274.
19. Aslam, Z.; Ahmed, F.; Almogren, A.; Shafiq, M.; Zuair, M.; Javaid, N. An Attention Guided Semi-Supervised Learning Mechanism to Detect Electricity Frauds in the Distribution Systems. *IEEE Access* **2020**, *8*, 221767–221782. [CrossRef]
20. Lu, X.; Zhou, Y.; Wang, Z.; Yi, Y.; Feng, L.; Wang, F. Knowledge Embedded Semi-Supervised Deep Learning for Detecting Non-Technical Losses in the Smart Grid. *Energies* **2019**, *12*, 3452. [CrossRef]
21. Li, J.; Wang, F. Non-Technical Loss Detection in Power Grids with Statistical Profile Images Based on Semi-Supervised Learning. *Sensors* **2020**, *20*, 236. [CrossRef]
22. Wu, R.; Wang, L.; Hu, T. AdaBoost-SVM for Electrical Theft Detection and GRNN for Stealing Time Periods Identification. In Proceedings of the IECON 2018-44th Annual Conference of the IEEE Industrial Electronics Society, Washington, DC, USA, 21–23 October 2018; pp. 3073–3078.

23. Kong, X.; Zhao, X.; Liu, C.; Li, Q.; Li, Y. Electricity theft detection in low-voltage stations based on similarity measure and dt-ksvm. *Int. J. Electr. Power Energy Syst.* **2021**, *125*, 106544. [CrossRef]
24. Buzau, M.M.; Tejedor-Aguilera, J.; Cruz-Romero, P.; Gómez-Expósito, A. Detection of Non-Technical Losses Using Smart Meter Data and Supervised Learning. *IEEE Trans. Smart Grid* **2019**, *10*, 2661–2670. [CrossRef]
25. Yan, Z.; Wen, H. Electricity Theft Detection Base on Extreme Gradient Boosting in AMI. *IEEE Trans. Instrum. Meas.* **2021**, *70*, 1–9. [CrossRef]
26. Razavi, R.; Gharipour, A.; Fleury, M.; Justice, A.I. A practical feature-engineering framework for electricity theft detection in smart grids. *Appl. Energy* **2019**, *238*, 481–494. [CrossRef]
27. Zheng, Z.; Yang, Y.; Niu, X.; Dai, H.-N.; Zhou, Y. Wide and Deep Convolutional Neural Networks for Electricity-Theft Detection to Secure Smart Grids. *IEEE Trans. Ind. Informat.* **2018**, *14*, 1606–1615. [CrossRef]
28. Hasan, M.N.; Toma, R.N.; Nahid, A.-A.; Islam, M.M.M.; Kim, J.-M. Electricity Theft Detection in Smart Grid Systems: A CNN-LSTM Based Approach. *Energies* **2019**, *12*, 3310. [CrossRef]
29. Zhai, D.; Pan, Y.; Li, P.; Li, G. Estimating the Vigilance of High-Speed Rail Drivers Using a Stacking Ensemble Learning Method. *IEEE Sensors J.* **2021**, *21*, 16826–16838. [CrossRef]
30. Tang, Y.; Gu, L.; Wang, L. Deep Stacking Network for Intrusion Detection. *Sensors* **2022**, *22*, 25. [CrossRef]
31. Zhao, R.; Mu, Y.; Zou, L.; Wen, X. A Hybrid Intrusion Detection System Based on Feature Selection and Weighted Stacking Classifier. *IEEE Access* **2022**, *10*, 71414–71426. [CrossRef]
32. Tan, R.; Zhang, W.; Chen, S. Decision-Making Method Based on Grey Relation Analysis and Trapezoidal Fuzzy Neutrosophic Numbers Under Double Incomplete Information and its Application in Typhoon Disaster Assessment. *IEEE Access* **2020**, *8*, 3606–3628. [CrossRef]
33. Takiddin, A.; Ismail, M.; Nabil, M.; Mahmoud, M.M.; Serpedin, E. Detecting Electricity Theft Cyber-Attacks in AMI Networks Using Deep Vector Embeddings. *IEEE Syst. J.* **2021**, *15*, 4189–4198. [CrossRef]
34. Seghouane, A.-K.; Shokouhi, N.; Koch, I. Sparse Principal Component Analysis with Preserved Sparsity Pattern. *IEEE Trans. Image Process.* **2019**, *28*, 3274–3285. [CrossRef]
35. Pavlyshenko, B. Using Stacking Approaches for Machine Learning Models. In Proceedings of the 2018 IEEE Second International Conference on Data Stream Mining & Processing (DSMP), Lviv, Ukraine, 21–25 August 2018; pp. 255–258.
36. Wang, X. Design and Implementation of College English Teaching Quality Evaluation System Based on Analytic Hierarchy Process. In Proceedings of the 2020 International Conference on Computers, Information Processing and Advanced Education (CIPAE), Ottawa, Canada, 16–18 October 2020; pp. 213–216.
37. Yin, J.; Han, L.; Ma, L.; Cai, H.; Li, H.; Li, J.; Sun, G. Evaluation of Terminal Signal Quality based on Entropy Weight Method. In Proceedings of the 2022 4th International Conference on Intelligent Control, Measurement and Signal Processing (ICMSP), Hangzhou, China, 8–10 July 2022; pp. 855–858.
38. Adeli, E.; Li, X.; Kwon, D.; Zhang, Y.; Pohl, K.M. Logistic Regression Confined by Cardinality-Constrained Sample and Feature Selection. *IEEE Trans. Pattern Anal. Mach. Intell.* **2020**, *42*, 1713–1728. [CrossRef] [PubMed]
39. Shuan, L.; Ying, H.; Xu, Y.; Ying, S.; Jin, W.; Qiang, Z. Electricity Theft Detection in Power Grids with Deep Learning and Random Forests. *J. Electr. Comput. Eng.* **2019**, *2019*, 1–12. [CrossRef]
40. Tang, M.; Zhao, Q.; Ding, S.X.; Wu, H.; Li, L.; Long, W.; Huang, B. An Improved LightGBM Algorithm for Online Fault Detection of Wind Turbine Gearboxes. *Energies* **2020**, *13*, 807. [CrossRef]
41. Xie, R.; Cui, Z.; Jia, M.; Wen, Y.; Hao, B. Testing Coverage Criteria for Deep Forests. In Proceedings of the 2019 6th International Conference on Dependable Systems and Their Applications (DSA), Harbin, China, 3–6 January 2020; pp. 513–514.
42. Vieira, J.; Duarte, R.P.; Neto, H.C. kNN-STUFF: kNN STreaming Unit for Fpgas. *IEEE Access* **2019**, *7*, 170864–170877. [CrossRef]
43. Wu, Y.; Liu, Y.; Li, N.; Wang, S. Hybrid Multi-objective Particle Swarm Optimization Algorithm Based on Particle Sorting. In Proceedings of the 2021 IEEE International Conference on Emergency Science and Information Technology (ICESIT), Chongqing, China, 22–24 November 2021; pp. 257–260.
44. Yao, D.; Wen, M.; Liang, X.; Fu, Z.; Zhang, K.; Yang, B. Energy Theft Detection with Energy Privacy Preservation in the Smart Grid. *IEEE Internet Things J.* **2019**, *6*, 7659–7669. [CrossRef]
45. Buzau, M.; Tejedor-Aguilera, J.; Cruz-Romero, P.; Gómez-Expósito, A. Hybrid Deep Neural Networks for Detection of Non-Technical Losses in Electricity Smart Meters. *IEEE Trans. Power Syst.* **2020**, *35*, 1254–1263. [CrossRef]

Article

Hybrid-Model-Based Digital Twin of the Drivetrain of a Wind Turbine and Its Application for Failure Synthetic Data Generation

Ainhoa Pujana [1,*], Miguel Esteras [1], Eugenio Perea [1], Erik Maqueda [1] and Philippe Calvez [2]

[1] TECNALIA, Basque Research and Technology Alliance (BRTA), Edificio 700, 48160 Derio, Spain
[2] ENGIE LAB CRIGEN, 1 Place Samuel De Champlain, 92400 Courbevoie, France
* Correspondence: ainhoa.pujana@tecnalia.com

Abstract: Computer modelling and digitalization are integral to the wind energy sector since they provide tools with which to improve the design and performance of wind turbines, and thus reduce both capital and operational costs. The massive sensor rollout and increase in big data processing capacity over the last decade has made data collection and analysis more efficient, allowing for the development and use of digital twins. This paper presents a methodology for developing a hybrid-model-based digital twin (DT) of a power conversion system of wind turbines. This DT allows knowledge to be acquired from real operation data while preserving physical design relationships, can generate synthetic data from events that never happened, and helps in the detection and classification of different failure conditions. Starting from an initial physics-based model of a wind turbine drivetrain, which is trained with real data, the proposed methodology has two major innovative outcomes. The first innovation aspect is the application of generative stochastic models coupled with a hybrid-model-based digital twin (DT) for the creation of synthetic failure data based on real anomalies observed in SCADA data. The second innovation aspect is the classification of failures based on machine learning techniques, that allows anomaly conditions to be identified in the operation of the wind turbine. Firstly, technique and methodology were contrasted and validated with operation data of a real wind farm owned by Engie, including labelled failure conditions. Although the selected use case technology is based on a double-fed induction generator (DFIG) and its corresponding partial-scale power converter, the methodology could be applied to other wind conversion technologies.

Keywords: wind turbine; digital twin; hybrid model; failure diagnosis; synthetic data generation; predictive maintenance

Citation: Pujana, A.; Esteras, M.; Perea, E.; Maqueda, E.; Calvez, P. Hybrid-Model-Based Digital Twin of the Drivetrain of a Wind Turbine and Its Application for Failure Synthetic Data Generation. *Energies* **2023**, *16*, 861. https://doi.org/10.3390/en16020861

Academic Editor: Francesco Castellani

Received: 17 November 2022
Revised: 30 December 2022
Accepted: 1 January 2023
Published: 12 January 2023

1. Introduction

In modern times, wind energy conversion is one of the most promising and reliable energy technologies. Europe already has 220 GW of wind capacity installed and there are plans to install an additional power of 105 GW over the next five years [1]. Actors involved in this energy source are continuously researching this technology with the aim of achieving the best levelized cost of energy (LCOE). According to WindEurope, operation and maintenance (O&M) expenses account for 25–35% of LCOE of wind turbines [2], where corrective maintenance is responsible for 30–60% of O&M costs [3]. The current potential of digitalization and artificial intelligence (AI) can greatly contribute to the increase in the energy production of wind farms, reducing unplanned interruptions, optimizing O&M, and extending the lifetime of the components.

Wind turbines systems can be classified depending on the type of generator, gearbox and power converter used. A double-fed induction generator (DFIG) with a multiple stage gearbox and a partial scale converter is a widely used technology [4]. In the DFIG topology [5], there is a direct connection between the stator windings and the constant frequency grid while the rotor winding connection to the grid is made through a pulse width modulation (PWM) power converter, using a set of slip rings. The power converters can control the rotor circuit current, frequency, and phase angle shifts [6]. This kind of induction generator can operate in a range of ±30% of synchronous speed, achieving a high energy yield, a power fluctuation reduction and the capability of controlling reactive power. A drawback of the DFIG is the inevitable need for slip rings.

A wind turbine is also equipped with a control system, which is responsible for assuring the correct operation of the wind turbine along its entire power curve and keeping the wind turbine within its normal operating range. Wind turbines contain electrical, mechanical, hydraulic, or pneumatic systems, and require sensors to monitor the variables that determine the required control action. The most common variables sensed in a control system are wind speed, rotor speed, active and reactive power, voltage, and the frequency of the wind turbine's connection point. Moreover, the control system is responsible for stopping the wind turbine if necessary. One control strategy is the pitch angle control [7], which is a good option for variable-speed operations in wind turbines generating more than 1 MW. Using this control, the blades can be correctly oriented with respect to the wind direction in order to avoid extremal values (too high or too low) of the power output. The pitch system is based on a hydraulic system, which requires a computer system or an electronically controlled electric motor.

There are several studies that analyse the critical failure modes of the wind turbine drive-train system, specifically the electric generator and power conversion system [8–10]. While identifying the sources of failure in the electric generator [11], the typologies of failures can be of different kind. Thermal failures can occur due to the effect that currents and overcurrents circulating through the windings have on the insulation and considering that a maximum temperature is withstood depending on the type of insulation and operating conditions. Electrical failures can also occur due to the peaks of voltage that can be applied to the conductor under normal operating conditions and in anomalous situations, such as surges coming from the converter. Environmental failures can be caused by environmental conditions that could degrade insulating material or create corrosion phenomena. Mechanical failures are mainly caused by vibrations. Finally, thermo-mechanical failures are caused by cyclic operating conditions with sudden or continuous variations in temperature, which have different effects depending on the cable material and its accessories (insulation, screens, etc.). The electric generator and the power converter have a greater impact on the reliability, failure rate, and unavailability of the wind turbine. Their failure rate is 15% per year for the electric generator and 6.8% for power converters of offshore wind farms [12,13]. These components are equipped with sensors (temperature, vibrations, electric parameters and others) and connected to the wind turbine supervisory control and data acquisition (SCADA) and condition-based monitoring (CBM) systems. Thus, a long historical real operation dataset exists for each turbine of a wind farm. Sometimes, this dataset includes recorded anomalies or failure in the operation of the turbine.

Data-driven models extract knowledge from real measurements that apply AI (artificial intelligence) techniques, which analyse large amounts of data to identify meaningful patterns in them. In the field of wind energy generation, there are several approaches for this type of model. For instance, the spectral analysis of current signals has been used for health monitoring of stator and rotor windings, as well as the main bearing of wind turbines [14]. In [15], a data-driven model is directly constructed with the objective of detecting and isolating sensor and actuator failures in wind turbines, while the study of [16] develops a hierarchical bank of negative selection algorithms (NSAs) to detect and isolate common failures in wind turbines. The study of [17] uses a data-driven failure diagnosis and isolation (FDI) method for wind turbines. It consists of the implementation of long short-term memory (LSTM) networks for residual generators. The decision-making process is made by applying a random forest algorithm. These FDI methods are designed using experimental and historical data generated both under normal and failure conditions; therefore, the availability of well-developed databases that include labelled anomaly/failure data is mandatory. The accuracy of data-driven methods is generally poor for cases not included in a training dataset. In addition, black box models (e.g., deep learning models) show a low explainability, making it difficult for domain experts to interpret results and gain the required trust to make decisions based on the output of the models.

As a solution to this main drawback of data-driven models, DTs that use physics-based models are developed to make the DT self-explanatory. The term "digital twin" can be defined as "a virtual representation of a real-life system or asset with the same behaviour". It allows system states to be calculated using integrated models and data, aiding the decision-making process over its life cycle from design to decommissioning. The concept of DT was first described in David Gelernter's 1991 book *Mirror Worlds* [18], and the term "digital twin" was first mentioned in a roadmap report developed by John Vickers (NASA) in 2010. The DT concept consists of two distinct parts: (1) the physics-based model representing the asset and (2) the connection of the model with the real asset. This connection refers to the information transferred (automatically or manually) from the asset to the DT and the information that could be transferred from the DT to the asset and the operator. In this way, a DT can accurately estimate an asset's condition.

A DT is based on mathematic models that represent physical phenomena, making it possible to understand the behaviour of the real asset in each moment. In addition, using this physics-based model, it is possible to create synthetic data for events that have never happened before, acquiring knowledge of the behaviour in some conditions that in other cases would not be possible. Data-driven models can identify and prevent events that were measured in the past. However, the training process of the data-driven algorithms, either non supervised or supervised, always relies on historical data. DTs, on the contrary, provide two new information sources: firstly, physics-based models can allow us to understand their real behaviour, and secondly, physical simulation enables the generation of synthetic data for potential new scenarios, such as potential anomalies or failure conditions. Moreover, hybrid models, considered to be a combination of physics-based models and data analytics, provide a powerful tool for diagnosis and prognosis [19]. Hybrid models developed with this purpose are a good basis for DT creation.

The main advantage of a DT design for a specific industrial setting is the potential to simulate realistic scenarios that are difficult or costly to create in the real system. These scenarios might be used for the prescriptive analysis of new operating conditions, or for testing extreme conditions and responses to anomalies or failures. The main challenge is to develop a simulation method that can be parametrized to output scenarios that differ from normal operation and, in some cases, to simulate conditions that have never been seen before in the real system. The authors of [20] describe four main approaches for the generation of simulated scenarios based on: (1) a simplified physical model; (2) a more complex DT design to model the specific properties of the real scenario; (3) a parametrized statistical generative model built upon prior knowledge of the relationships between variables; and (4) generative models trained with existing real data distribution.

The methodology proposed in this paper brings together approaches 2 and 4 to develop a hybrid digital twin that combines physics-based models and data-driven models to match a specific operation context, both in normal and extreme or failure conditions. In addition, the DT preserves the constrains, significance and explainability of a physical model, overcoming some of the main limitations of a purely statistical generative model (i.e., generative adversarial networks). The physics-based model for the drivetrain of a wind turbine is developed using MATLAB Simulink R2020b.

The paper is organized as follows: Section 1 describes the developed technical approaches and the literature review related to such technical approaches, as well the problems of using data-driven approaches in comparison with hybrid models. Section 2 explains the proposed methodology for developing a hybrid-model-based digital twin and the advantages of combining both physics-based and data-driven models. Moreover, this section describes the principles of synthetic data generation and how such principles can be applied to failure data generation. In Section 3, this methodology is concretely applied to a use case: the drivetrain of a 1.5 MW wind turbine with DFIG technology. Section 4 contains the conclusions and perspectives of future research.

2. Methodology for a Hybrid Model Creation, Synthetic Failure Data Generation and Failure Classification Applied to a Digital Twin

DT development involves several technical tasks combining domain-specific knowledge and data analytics skills. First, the equipment or system deterministic model in normality conditions (so-called normality model) must be generated (e.g., by simulation model). This process includes the representative modelling of underlying physical phenomena and the rigorous selection of design parameters. Then, the constructed model must be validated using real data in non-failure conditions and optimizing certain model parameters values to increase the model accuracy and representativeness against the real equipment behaviour.

In addition, a DT conceived for failure conditions diagnosis includes a suite of physics-based models able to simulate different anomaly or failure scenarios. These failure models might be used for a cause–effect analysis and to establish condition indicators (CI) and they constitute an excellent basis for real failure conditions synthetic data generation [21]. Finally, machine learning (ML) classification techniques (supervised or non-supervised) might be applied for the diagnosis or early detection of failures. The implementation of all these models and algorithms in a digital platform and their online use constitute a complete DT for anomaly/failure diagnosis.

This chapter describes and analyses the methodology for the development and use of an equipment or system DT based on hybrid models for failure classification, making use of a normality hybrid model and a synthetic data generation process. Figure 1 summarizes the whole methodology, and each key component is explained in the following chapters.

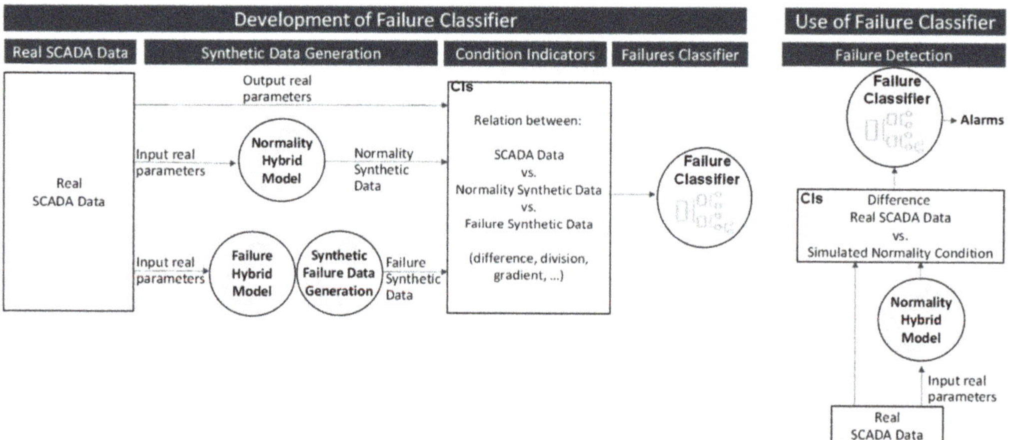

Figure 1. Methodology illustration for the creation of a hybrid-model-based DT.

2.1. Normality Hybrid Model

The normality hybrid model of the DT is composed of a physics-based model trained with real operation SCADA data in normality conditions.

The paper considers the drivetrain of a wind turbine with DFIG technology as a reference use case in which the proposed DT development methodology is illustrated and applied. Figure 2 shows how the physics-based model is divided in two modules that could be used either coupled together or separately, depending on the available operational data. The first module represents the conversion from kinetic energy from the wind to mechanical power, taking the real values of the wind speed measured at the turbine and the pitch angle of the blades as inputs. The second module represents the electro-mechanical conversion. It takes the mechanical torque in the shaft of the DFIG as the input and the generated electric power and its related signals, such as phase currents and voltages or electromagnetic torque, are the outputs. Moreover, this second module includes a power converter and control system that enables the optimal operation of the drivetrain.

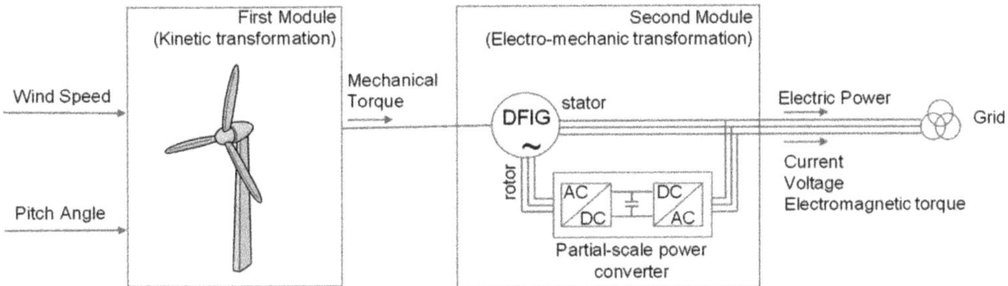

Figure 2. Physics-based model of the power conversion drivetrain of a wind turbine.

The physics-based model is constructed considering the system design parameters. Depending on the nature of the equipment it may be difficult to obtain the complete set of design parameters. In this case, estimations are required, which may impact model performance. Finally, the physics-based model is trained using real operation SCADA data (Figure 3). Training consists of optimizing the values of certain independent design parameters whose exact values are estimated between given realistic intervals.

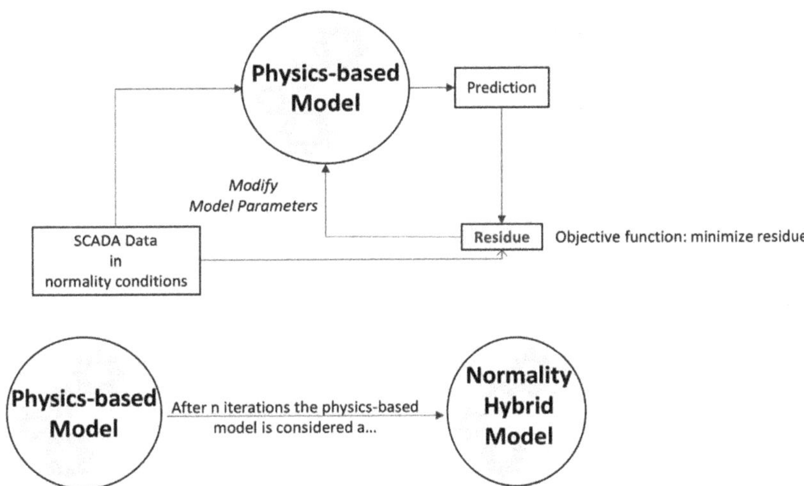

Figure 3. Training of the physics-based model and obtention of the normality hybrid model.

The objective function of the training process is the minimization of "residue" defined as the difference between the physics-based model output (prediction) and the SCADA real operation data (e.g., output power) for the given real inputs (e.g., wind speed or torque). The resulting calibrated physical model is known as the normality hybrid model.

2.2. Failure Hybrid Model

Once the normality hybrid model is constructed, it can be extended or adapted to include anomaly or failure situations. This new model is called a failure model. Following the same process used in the normality hybrid model, this model is trained using the operation real SCADA data. Similarly, calibration consists of optimizing the values of certain independent design parameters that represent failure, whose exact values are estimated between given realistic intervals.

This resulting new model is also trained with historical and actual operational data of both normal and failure operation. This is achieved using real failure operation data inputs, which are fed to the failure models. In other words, when the normality hybrid model is adapted to represent a failure and trained with failure data (data representing failure operation), the normality hybrid model becomes a failure hybrid model. Feeding the failure models with failure data enables the values of the failure model parameters that define the failure models to be calibrated. The selected values of these failure model parameters are obtained by minimizing the difference between the prediction obtained by the failure model using failure operation data inputs and their corresponding well-known real operation data failure outputs. As a result, the so-called failure hybrid model of the power conversion system (drivetrain) of a wind turbine is obtained, which considers both data of the drivetrain in normal operation and in failure operation.

In this case, the overheating of the DFIG stator winding is studied. For this scenario, a thermal model is added to the normality hybrid model (Figure 4).

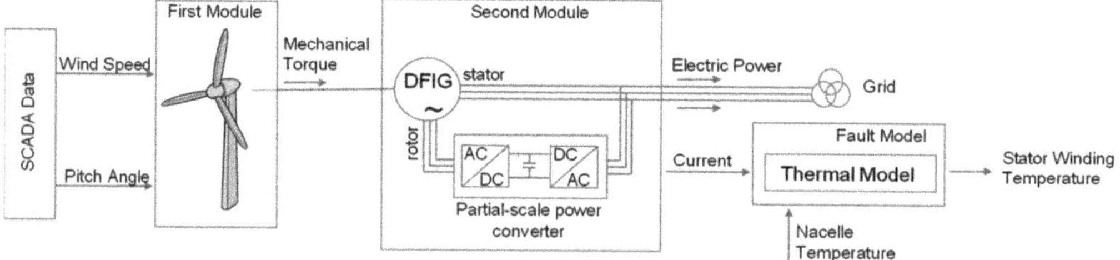

Figure 4. Failure hybrid model with a specific thermal failure model.

This thermal model takes as input the real values of the nacelle temperature and the stator phase currents. These values of these stator currents can be estimated by the normality hybrid model or any other value that can be useful for testing the thermal behaviour of DFIG stator windings. The obtained predicted output corresponds to the temperature of the DFIG stator winding.

2.3. *Failure Synthetic Data Generation*

The methodology analysed in the article has a fundamental contribution in the generation of synthetic data. The generation of synthetic data is a key point because it allows immediate availability of operation data (either normality or failure data), that are difficult to obtain from simple observation of the reality. In addition, the training of classification models for failure prognosis is much enriching using a broad and balanced dataset that represents a variability of behaviour.

Ref. [22] proposes GANs for the generation of synthetic data for wind turbine failure diagnosis research. This article proposes a method to generate synthetic data using the hybrid model and a statistical process. The statistical process characterizes the probability distributions of the occurrence of normal and failure operating scenarios.

The generation of synthetic scenarios in a DT is often deterministic; therefore, the given input data (i.e., wind speed, nacelle temperature and blade pitch angle) always calculate the same output data (i.e., active power, winding temperature, etc.). This process does not consider the variance present in the real data due to factors not modelled by the DT. Hence, the DT does not have the ability to interpolate within the space of the training data and cannot generate truly new scenarios, nor can it include the full extent of the variability observed in the data. In the case of the generation of normal condition scenarios, this determinism is compensated by the amount of training data in such conditions. It is reasonable to assume that these data include a comprehensive range of conditions that represent the entire feature space.

However, this might not be the case for the generation of failure conditions. Although the failure hybrid model has been calibrated to simulate the instances belonging to this type of conditions present in the training SCADA data, this does not guarantee that these instances are a representation of the entire anomalous feature space. In fact, the frequency of anomalous conditions and failures is relatively low in SCADA data, and often these instances are not annotated (labelled). Hence, relaying merely on a deterministic model to generate synthetic failure scenarios would provide a narrow data sample constrain to patterns already seen before.

To resolve this limitation, the DT can incorporate stochastic failure models for the generation of failure scenarios. Each of these models can generate an unlimited number of synthetic failure scenarios for a particular failure type based on real observations in SCADA data.

The corresponding models are trained to approximate the distributions of the variables that define a failure. In addition, some failures cannot be considered instantaneous, but as a pattern in time that leads to a malfunction, a safety stop or a break. This is especially important if synthetic generated failures are to be used to train models that can produce early warnings before a failure is likely to occur.

Both the join probability distribution of the operating variables prior to and during a failure and their physical constrains are initially defined by domain knowledge and can then be updated with observations from real SCADA data. The generation of new failure scenarios is based on random sampling of these probability distribution. Hence, the synthetic scenarios generated by the model are based on real SCADA observations but are not identical to any of those. The process for the synthetic failure data generation of Figure 1 is detailed in Figure 5. It consists of two steps: an observation step and a synthetic data generation step. The observation step aims to identify the probability density function (PDF) that characterizes the failure scenario occurrence. For this, SCADA data are filtered to identify scenarios that correspond to a failure type f_k, where k is part of a set of failures K modelled by the DT, such that $k \in K$. A failure scenario is defined by a set of fixed physical constrains defined by domain knowledge and a set of parameters (condition indicators) to be tuned in function of the observed features in failure scenarios from SCADA data.

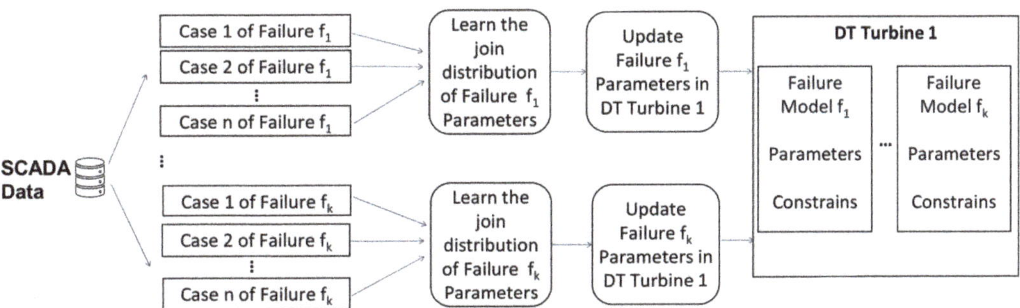

Figure 5. Observation process for failures.

The PDFs of the parameters are learnt from the observed instances in the SCADA data. These instances might be exclusively sourced from a single turbine or, in case of an insufficient number, they can be sourced from different turbines that share some design and operations characteristics. The decision to include instances from more than one turbine should be made on the basis of turbine similarity and the variability of failure parameters, which depends on operation and design characteristics. The distribution of most parameters might be approximated by a normal PDF with the required precision.

However, other distributions might need to be considered for certain parameters. In the case of having access to SCADA data with several instances of a given failure for more than one turbine, a hierarchical parameter modelling might provide a better balance between accuracy and generalization. The learnt PDFs of the parameters are used to update the prior parameter distributions of the corresponding failure model. The data generation process step consists of generating data sets for normality and failure scenarios. As shown in Figure 6, the normality scenario data sets are generated either by running the normality hybrid model or selecting those SCADA data labelled as normal data.

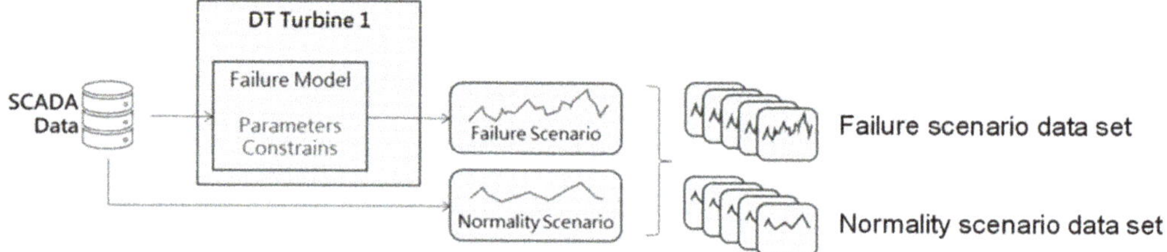

Figure 6. DT generative failure models.

The failure scenario data sets are stochastically generated following the observed and identified PDF, then running and obtaining the results from the failure hybrid model.

2.4. *Potential Application of the Hybrid Models Conforming the Digital Twin*

The development of data-driven algorithms for diagnosing normality or failure conditions is a complex task that involves: (i) defining the condition indicators (CIs), (ii) labelling normality and failure operation data, (iii) conceptualization of the classification model, (iv) validation of the model (e.g., number of false positives and negatives), and (v) evaluation of the generalization capacity of the model analysing whether it is representative for a set of machines. The DT can add value to this endeavour by providing additional synthetic data to strengthen the dataset.

Figure 7 shows a proposed schema of a supervised classifier training process for failure diagnosis where the explained models in the previous sections are leveraged. The classifier is trained with a labelled dataset composed of real SCADA data, augmented with synthetic data generated via the process described in the previous section.

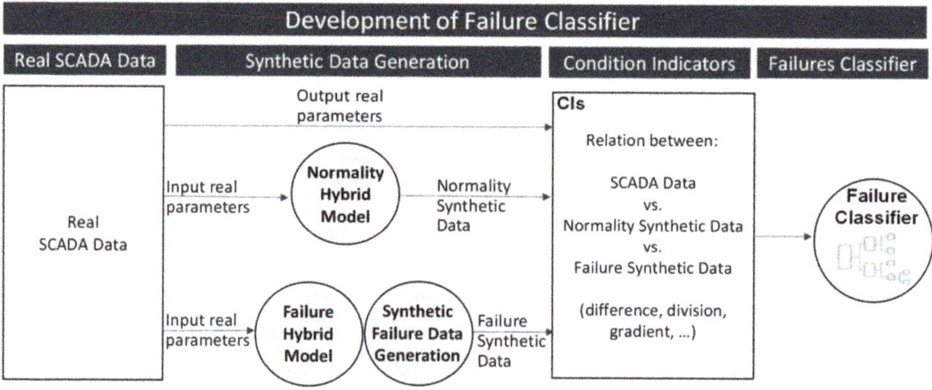

Figure 7. Supervised classifier training scheme.

In addition, the normality hybrid model is used as a baseline to create new CIs that may improve the accuracy of the classifier. These CIs are calculated by comparing real operation

SCADA data with respect to synthetic failure data and/or normality data generated by the normality hybrid model.

Finally, Figure 8 shows the execution phase, where CIs are created by comparing real SCADA data with the data simulated by the normality hybrid model. When the values of these CIs meet certain criteria detected by the classifier, an early alarm is generated.

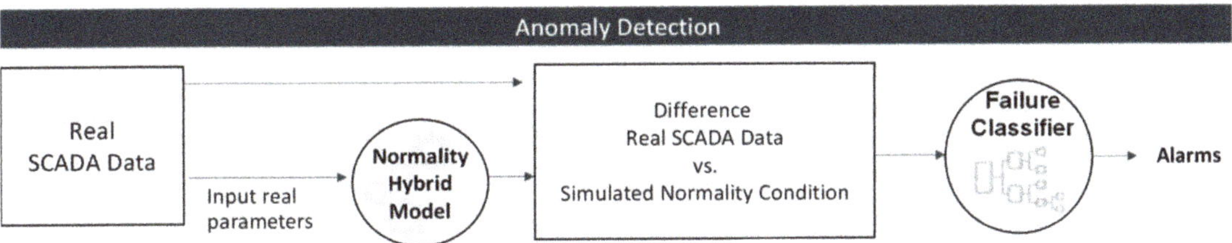

Figure 8. Execution phase of the developed classifier for anomaly diagnosis.

3. Results of Application of the Methodology to a Use Case: 1.5 MW DFIG Wind Turbine

The methodology described in previous section was applied and validated with real SCADA data from a wind turbine in operation owned by Engie. The drivetrain of this wind turbine comprises a 1.5 MW DFIG and its corresponding back-to-back power converter.

Three years of real operational data were organized and preprocessed before use. During the data exploration and pre-processing of SCADA data, relationships between physic parameters were analysed, in order to detect possible outliers, which were removed.

Once the initial data analysis was carried out, the physical model of the power conversion was developed in Simulink-Matlab R2020b (Figure 9). Information on the design parameters of both the generator and power converter was used as a basis for constructing the model. However, some other values were calculated or estimated due to the lack of information. Wind speed and pitch angle are the input parameters needed to operate the model. The result is the generated electric power, currents, and voltages, among others.

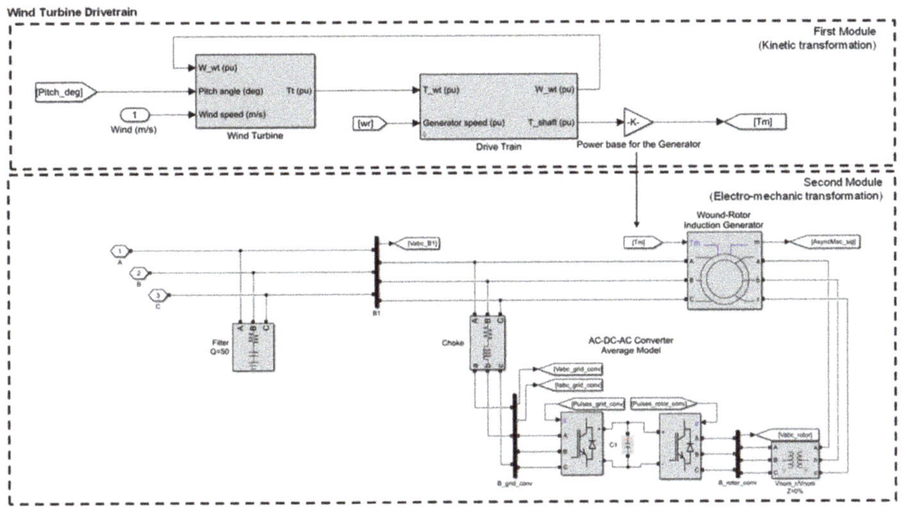

Figure 9. Wind turbine drivetrain physics-based model representation in Matlab-Simulink.

The DFIG block implements a three-phase wound rotor asynchronous machine, operating in the generator mode. It uses a fourth-order state-space model to represent the electrical part of the machine, whereas the mechanical part is represented by a second-order system. As can be seen in the equations contained in Table 1, all the electrical parameters are referred to in the stator. All the rotor and stator parameters are expressed in the arbitrary two-axis reference dq frame.

Table 1. Equivalent circuits and equations involved in a DFIG conversion.

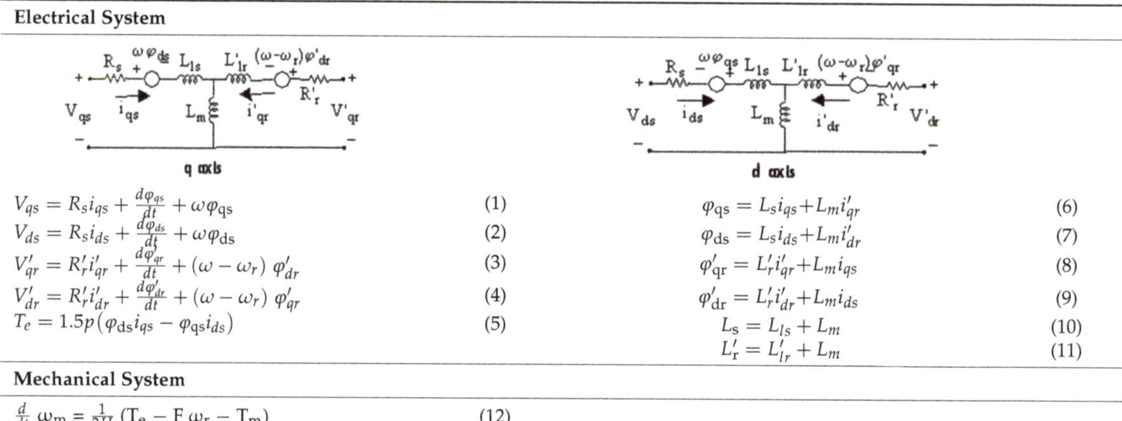

Electrical System

$$V_{qs} = R_s i_{qs} + \frac{d\varphi_{qs}}{dt} + \omega\varphi_{qs} \quad (1)$$
$$V_{ds} = R_s i_{ds} + \frac{d\varphi_{ds}}{dt} + \omega\varphi_{ds} \quad (2)$$
$$V'_{qr} = R'_r i'_{qr} + \frac{d\varphi'_{qr}}{dt} + (\omega - \omega_r)\,\varphi'_{dr} \quad (3)$$
$$V'_{dr} = R'_r i'_{dr} + \frac{d\varphi'_{dr}}{dt} + (\omega - \omega_r)\,\varphi'_{qr} \quad (4)$$
$$T_e = 1.5p(\varphi_{ds} i_{qs} - \varphi_{qs} i_{ds}) \quad (5)$$
$$\varphi_{qs} = L_s i_{qs} + L_m i'_{qr} \quad (6)$$
$$\varphi_{ds} = L_s i_{ds} + L_m i'_{dr} \quad (7)$$
$$\varphi'_{qr} = L'_r i'_{qr} + L_m i_{qs} \quad (8)$$
$$\varphi'_{dr} = L'_r i'_{dr} + L_m i_{ds} \quad (9)$$
$$L_s = L_{ls} + L_m \quad (10)$$
$$L'_r = L'_{lr} + L_m \quad (11)$$

Mechanical System

$$\frac{d}{dt}\,\omega_m = \frac{1}{2H}\,(T_e - F\,\omega_r - T_m) \quad (12)$$
$$\frac{d}{dt}\,\Theta_m = \omega_m \quad (13)$$

The parameters involved in the resolution of DFIG conversion equations are those indicated in Table 2.

Table 2. Parameters involved in the DFIG operation.

Parameters	Definition
R_s, L_{ls}	Stator resistance and leakage inductance
L_m	Magnetizing inductance
L_s	Total stator inductance
V_{qs}, i_{qs}	q axis stator voltage and current
V_{ds}, i_{ds}	d axis stator voltage and current
ϕ_{qs}, ϕ_{ds}	Stator q and d axis fluxes
p	Number of pole pairs
ω	Reference frame angular velocity
ω_m	Mechanical angular velocity
ω_r	Electrical angular velocity ($\omega_m \times p$)
Θ_m	Mechanical rotor angular position ($\Theta_m \times p$)
Θ_r	Electrical rotor angular position ($\Theta_m \times p$)
T_e	Electromagnetic torque
T_m	Shaft mechanical torque
J	Combined rotor and load inertia coefficient (set to infinite to simulate locked rotor)
H	Combined rotor and load inertia constant (set to infinite to simulate locked rotor)
F	Combined rotor and load viscous friction coefficient
L'_r	Total rotor inductance
R'_r, L'_{lr}	Rotor resistance and leakage inductance
V'_{qr}, i'_{qr}	q axis rotor voltage and current
V'_{dr}, I $would_r$	d axis rotor voltage and current
ϕ'_{qr}, ϕ'_{dr}	Rotor q and d axis fluxes

3.1. Normality Hybrid Model of the Use Case

The initial parameters of the physics-based model are an assumption of the true parameters controlling the operation of a given turbine. Nevertheless, the true value of these parameters can be estimated using an optimization algorithm. The algorithm aims to find the combination of parameter values that minimize the difference between the output of the physics-based model and the measured SCADA data. In this case, the parameters are tuned (or calibrated) using a surrogated optimization algorithm (surrogateopt) in Matlab [23]. This optimization algorithm is a global solver specially indicated for cases where the objective function is computationally expensive. The algorithm searches for a global minimum of a cost function $\min_x f(x)$ with multivariate input variable x subject to linear and non-linear constrains, and some finite bounds. The resulting objective function can be non-convex and non-smooth. The algorithm starts by learning a surrogate model of the function considered as objective, using the interpolation of radial basis function through random evaluations of the objective function within the given bounds. In the next phase, a merit function is minimized by approximating the minimization of the objective function. This merit function f_m is based on a weighted combination of the evaluation of the surrogate model calculated in the previous phase, and the distance between the points sampled from the objective function.

$$f_m(x) = wS(x) + (1 - w)D(x) \tag{14}$$

$$S(x) = \frac{s(x) - s_{min}}{s_{max} - s_{min}} \tag{15}$$

$$D(x) = \frac{d_{max} - d(x)}{d_{max} - d_{min}} \tag{16}$$

where $S(x)$ is a scaled surrogated output and $D(x)$ is a scale distance between points evaluated by the objective function. This distance reflects the uncertainty in the estimations of the surrogate model. The minimization of the merit function, $\min_x f_m(x)$, is performed using a random search. The obtained global minimum is then evaluated by the objective function and the result used to update the surrogate model. Now the minimization of the merit function is calculated using the updated model. This process continues for a given number of iterations or until a point is found for which the objective function is below a threshold.

In the case of the drivetrain of the wind turbine, the objective function is defined as the mean absolute percentage error (MAPE) between the active power estimated by the physics-based model and the active power measured by the SCADA system.

$$\text{MAPE} = \frac{100\%}{n} \sum_{i=1}^{n} \left| \frac{PkW_i^{sim} - PkW_i^{real}}{PkW_i^{real}} \right| \tag{17}$$

Thirteen parameters are involved in the optimization process: four parameters associated with electro-mechanic conversion (electric generator, power converter and wind turbine control), three parameters related to aero-dynamical conversion, three parameters of the control strategy, and finally, three parameters associated with the mechanical drivetrain (Table 3).

The calibration was made in two steps: in the first step, six variables were considered, while in the second step, five more variables were added. Table 4 shows both the initial values defined for each parameter (design value), as well as the values adopted after second calibration (calibrated value).

Table 3. Parameters involved in the optimization process.

Electric Generator	Power Converter	Control	Mechanical Drivetrain
Stator winding resistance	Power converter grid-side coupling resistance	DC bus voltage regulator gains	Wind turbine inertia constant
Rotor winding resistance	Power converter grid-side coupling inductance	Speed regulator gains	Shaft mutual damping
Generator inertia constant	Converter line filter capacitor	Wind speed at nominal speed and at Cp max	Shaft spring constant
Generator friction factor			

Table 4. Design and calibrated values of parameters involved in the optimization process.

Parameters to Be Calibrated	Design Values	Calibrated Value
Stator winding resistance (pu)	0.016	0.0036
Rotor winding resistance (pu)	0.023	0.001
Generator inertia constant	0.685	0.1
Generator friction factor	0.01	0.01
Power converter grid-side coupling resistance (pu)	0.03	0.0232
Power converter grid-side coupling inductance (pu)	0.3	0.4811
Converter line filter capacitor (VAr)	120,000	89,200
DC bus voltage regulator gains	400, 8	323, 6.36
Speed regulator gains	0.6, 3	0.69, 2.67
Wind speed at nominal speed and at Cp max (m/s)	11	10
Wind turbine inertia constant (s)	4.32	2
Shaft mutual damping	1	1
Shaft spring constant	1.5	0.5

The new values of the calibrated parameters are established, always keeping their physical sense. In fact, an interval with a lower and upper threshold was established for each parameter during the optimization process.

As a result, the mean absolute percentage error (MAPE) between the real active power measured in the SCADA and the value obtained in the simulation using the calibrated models improved from 15% to 2.4% (Figure 10).

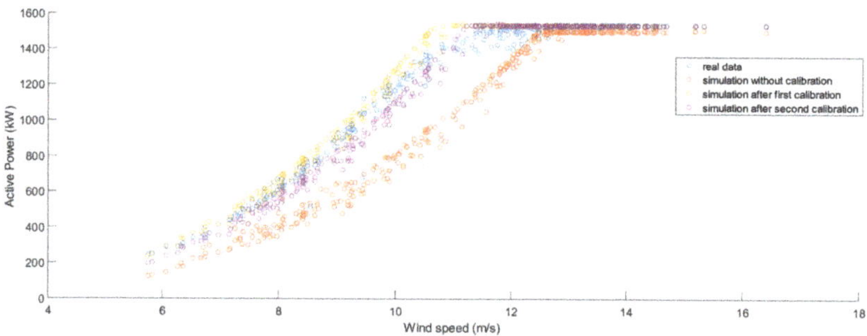

Figure 10. Generated active power vs. wind speed.

3.2. Failure Hybrid Model of the Use Case

Once the physic model was calibrated, it was used to simulate the failure conditions. In this use case, the overtemperature in the stator winding was analysed. A thermal circuit was added to the already developed normality hybrid model in Simulink to estimate the temperatures in each phase of the stator winding. It must be considered that the isolation class of the stator winding is a Class F, meaning that it is designed to withstand temperatures of up to 155 °C. As shown in Figure 11, this thermal circuit takes into account heat transference generated by the stator currents considering the conduction (between the winding of each one of the three stator phases) and convection (between the winding of each one of the three stator phases, between each stator winding and the environment and between each stator winding and the rotor). The values of radiation were neglected.

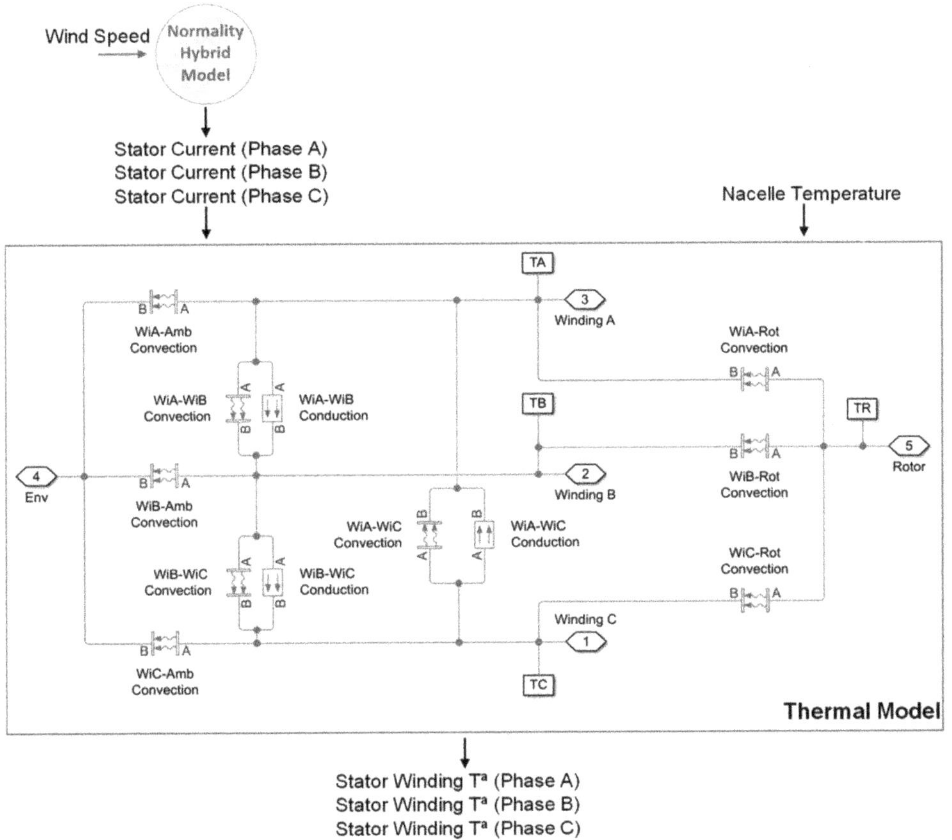

Figure 11. Thermal circuit of stator winding.

Conductive heat transfer blocks model heat transfer in the thermal network by conduction through a layer of material. The rate of heat transfer is governed by Fourier's law (18) and is proportional to the temperature difference, material thermal conductivity, area normal to the heat flow direction, and inversely proportional to the layer thickness.

$$Q_{cond} = \frac{k}{s} A\, dT \tag{18}$$

Convective heat transfer blocks model heat transfer in a thermal network by convection due to fluid motion (in this case, the air). The rate of heat transfer (19) is proportional to the temperature difference, heat transfer coefficient and surface area in contact with the fluid.

$$Q_{conv} = hc\, A\, dT \tag{19}$$

The inputs that feed the thermal model are the stator currents and the room temperature where the electric generator is installed (in this case the temperature of the nacelle), while the outputs are the temperatures of each phase of the stator winding.

In the real data made available during this study, there are five anomaly cases labelled as overtemperature in the stator winding (Figure 12).

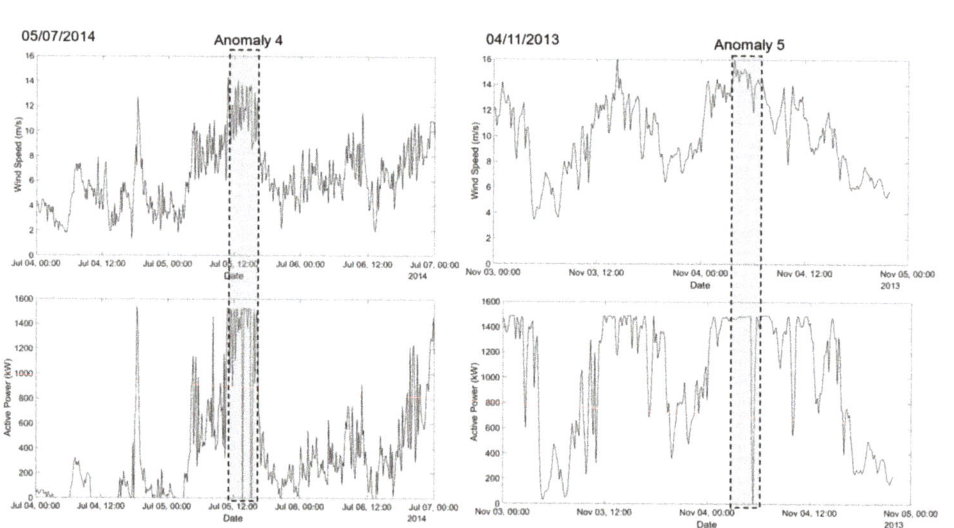

Figure 12. Five labelled anomaly cases of overcurrent during real operation (wind speed and active power signals).

The failure modelling was validated using data during these five anomaly cases, obtaining results for the estimated stator winding temperatures, as shown in Figure 13, compared with the real SCADA winding temperature.

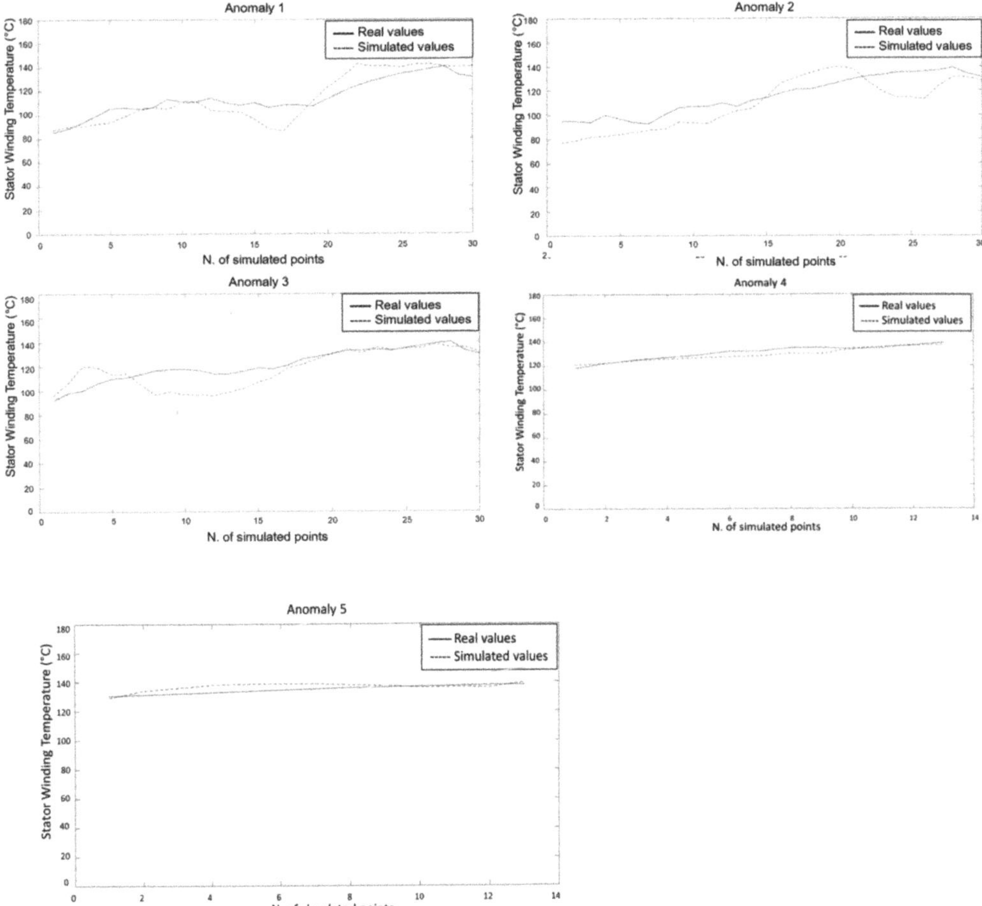

Figure 13. Temperature values during overtemperature failure-labelled cases.

The MAPE between the real stator winding temperature measured in the SCADA and the value obtained in the simulation using the calibrated model has a value of 11%, with a maximum percentage error of 16% in the worst scenario. This value still has room for improvement if more accurate design data become available for the thermal model.

3.3. Synthetic Failure Data Generation in the Use Case

A failure model for stator winding overheating was trained with real data from five labelled failures. For this failure mode, four parameters (CIs) were identified: failure or anomaly duration, ambient temperature, nacelle temperature, and wind speed.

The failure duration and ambient temperature are assumed to be uniform during the whole duration of the failure. The distribution of these values in the training data is approximated with a kernel density function (KDE) with a Gaussian kernel (Figure 14). Continuous line represents the probability density functions of the duration and ambient temperature observed in the failure/anomaly instances from the real SCADA, while cross symbols represent real observations This technique, compared with density estimation by histogram, creates a smooth PDF that does not depend on the choice of binning. Instead, a Gaussian component is fitted to each data point. The Gaussian kernel is defined by the function:

$$K(x;h) \propto \exp\left(-\frac{x^2}{2h^2}\right) \tag{20}$$

where the density function estimated at point x of a univariate distribution is:

$$\hat{f}(x;h) = n^{-1}\sum_{i=1}^{n} K(x - x_i;h) \tag{21}$$

where $(x_1,\ x_2,\ \ldots,\ x_n)$ are independent and identically distributed random samples from such distribution. The bandwidth h is a smoothing parameter that controls the balance between variance and bias in the resulting density function. The resulting Gaussian mixture is a non-parametric estimator of the probability density function able to represent the uncertainty present in a small data sample. In addition, a domain expert can intuitively control the estimator with a bandwidth parameter based on a descriptive analysis of SCADA data and physical properties of the system.

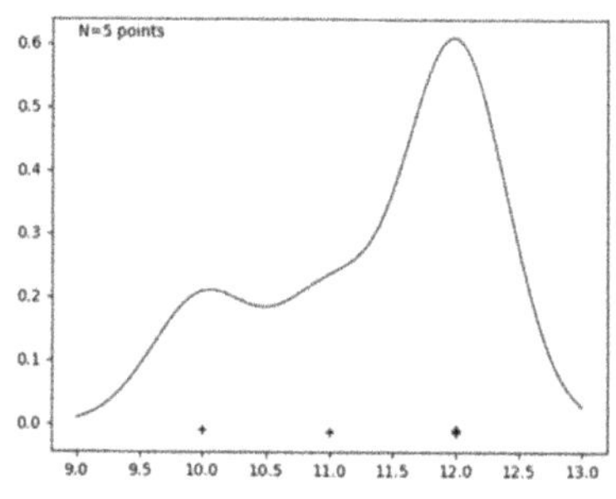

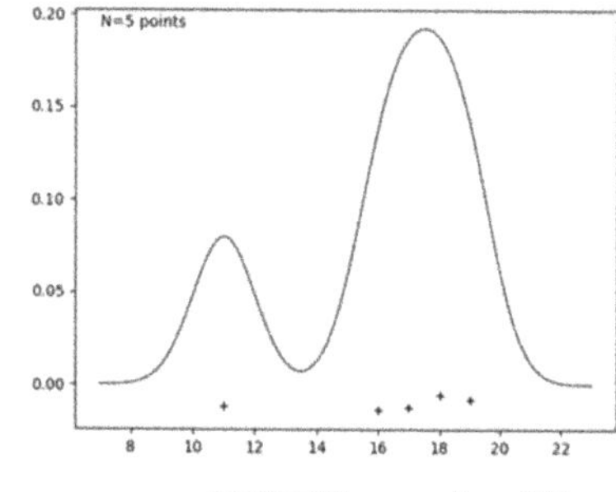

Figure 14. Probability density functions (continuous line) of the duration and ambient temperature observed in the failure/anomaly instances from the real SCADA. Cross symbol represents real observations.

The PDF of the wind speed and nacelle temperature variables are dependent on the relative time within a given failure or anomaly. Hence, a generative model aims to learn a PDF from which to sample a time series of a given variable, not simply a single value. Such a function can be approximated by recursively fitting an ordinary least squares (OLS) model to the transition between each time point. In this case, the resulting marginal probability distribution at a given point in time is conditional to the value at the previous time point. The statistical model of the predicted value is:

$$X_{t1} = X_{t0}\beta + \varepsilon \tag{22}$$

Additionally, the estimation error ε is assumed to have a normal distribution such that:

$$\varepsilon|X_{t0} \sim N\left(0,\ \sigma^2 I\right) \tag{23}$$

where σ^2 is a positive common variance for the elements of the error vector (assuming homoscedasticity) and I is the identity matrix.

The generation of random samples starts by the sampling an unconditional seed at time 0. This seed is randomly sampled from a distribution learnt from the training values at time 0. The distribution is approximated by KDE as seen above for the case of ambient temperature. The next data point in the time series, X_{t1}, is sampled from the distribution of ε around the prediction mean value $X_{t0}\beta$. This process iterates for each data point the requested time. Finally, synthetic failure patterns are randomly generated using the learnt statistical distributions (Figure 15) and are fed as inputs into the developed DT.

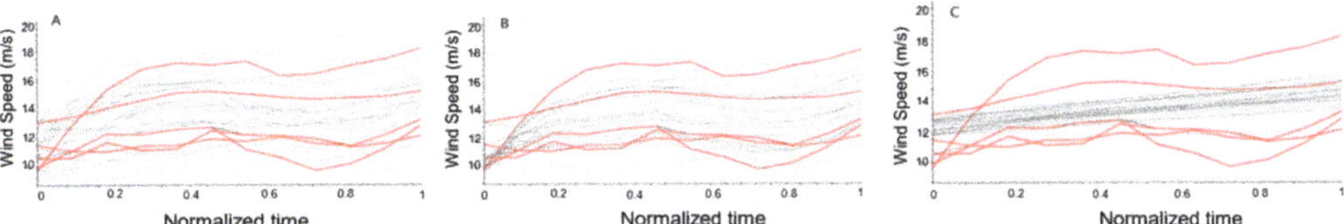

Figure 15. Generation of random patterns (in grey) of wind speed based on real SCADA data (in red).

The DT generates the rest of failure synthetic measurements (e.g., stator winding temperature, and generator output current,) creating a multivariate synthetic failure scenario (Figure 16).

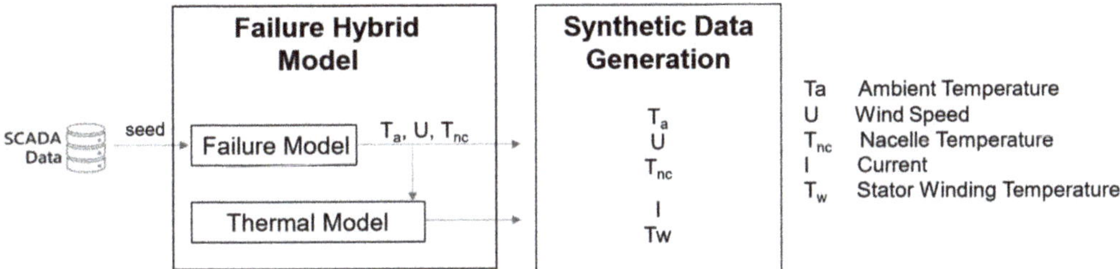

Figure 16. Multivariate synthetic failure pattern formed by the output of the data-driven stochastic model and the deterministic functions of the DT.

Figure 17 shows both the synthetically generated stator winding temperature values (in grey), and the stator winding real values measured by the SCADA system (in red). It can be noted that most of the synthetically generated data are similar to the real SCADA data. However, few of the synthetically generated data significantly differ from real data due to the starting seed value.

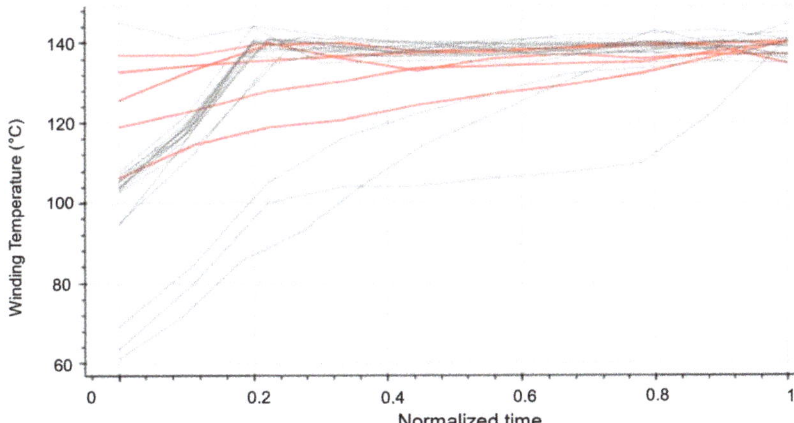

Figure 17. Stator winding temperature calculated by the DT thermal model from synthetic input variables (in grey). Stator winding temperature as measured by the SCADA system (in red).

4. Conclusions and Next Steps

This paper proposes an approach for creating a hybrid model-based digital twin that combines the benefits of physics-based models with advanced data analytics techniques.

This study has two main innovation outcomes. On the one hand, a process is established to generate synthetic failure data based on real data leveraging different statistical techniques. On the other hand, the process of failure classification based on machine learning techniques, allows anomaly conditions to be identified in the operation of the wind turbine. These two innovations can provide solutions for the main limitations of current digital twin approaches regarding accuracy, explainability, and the lack of sufficient training data.

The synthetic failure data generation process was validated using real operational data from a 1.5 MW power double-fed induction generator wind farm owned by Engie. In more detail, this has been applied to a specific failure (or anomaly) mode, namely the stator winding overtemperature. The obtained results are satisfactory, although further research is necessary. One of the limitations found in current research is the difficulty in achieving detailed labelled failure information.

In future studies, the authors foresee the following research lines. It is envisaged that a developed methodology for failure diagnosis, leveraging non-supervised and supervised machine learning algorithms, could be applied, as explained in Section 2.4. The results of this research could form the basis for future publications, which will likely be derived from the methodology of this article. These algorithms will be trained using real operational data augmented with synthetic failure data generated using this methodology. Furthermore, the authors plan to assess the generalization capacity of the proposed approach, validating it with additional failure modes and other drivetrain technologies (i.e., permanent magnets). Equally, the developed hybrid models might be further improved by applying state-of-the-art deep learning techniques. Finally, the scalability of the proposed solution should be assessed by implementing and validating it in an online real-time scenario.

5. Patents

The work reported in this manuscript is associated with a patent with application number EP22382724.7.

Author Contributions: Conceptualization, A.P., E.P. and E.M.; methodology, A.P., E.P. and E.M.; software, A.P. and M.E.; validation, A.P., M.E.; investigation, A.P., E.P. and E.M.; resources, P.C.; writing—original draft preparation, A.P.; writing—review and editing, A.P., M.E., E.P., E.M. and P.C.; visualization, P.C.; supervision, P.C.; project administration, E.M.; funding acquisition, A.P., E.P. and E.M. All authors have read and agreed to the published version of the manuscript.

Funding: This research was funded by the European Union Horizon 2020 Programme, grant number 872592 (PLATOON-2020), and the Elkartek Programme of Eusko Jaurlaritza KK-2018/00096 (VIRTUAL).

Data Availability Statement: Data used in this research are not publicly available due to Engie Green ownership: https://digital.engie.com/en/solutions/darwin (accessed on 4 October 2022). They might be available from the author Phillipe Calvez upon request.

Conflicts of Interest: The authors declare no conflict of interest.

Glossary

AI	Artificial Intelligence
CBM	Condition-Based Monitoring
CI	Condition Indicator
DFIG	Double Fed Induction Generator
DT	Digital Twin
FDI	failure diagnosis and isolation

GAN	Generative Adversarial Networks
KDE	Kernel Density Function
LCOE	Levelized Cost of Energy
LSTM	Long Short-Term Memory
MAPE	Mean Absolute Percentage Error
ML	Machine Learning
NSA	Negative Selection Algorithm
OLS	Ordinary Least Squares
O&M	Operation and Maintenance
PDF	Probability Density Function
PWM	Pulse Width Modulation
SCADA	Supervisory Control Additionally, Data Acquisition

References

1. Wind Energy in Europe: 2021 Statistics and the Outlook for 2022–2026. Wind Europe. Available online: https://windeurope.org/intelligence-platform/product/wind-energy-in-europe-2021-statistics-and-the-outlook-for-2022-2026/ (accessed on 4 October 2022).
2. Wind Energy Digitalisation towards 2030. Cost Reduction, Better Performance, Safer Operations. Published in November 2021. Available online: https://windeurope.org/intelligence-platform/product/wind-energy-digitalisation-towards-2030/ (accessed on 4 October 2022).
3. Hansen, A.D. Wind Energy Engineering: A Handbook for Onshore and Offshore Wind Turbines. In *Wind Turbine Technologies*; Academic Press: Cambridge, MA, USA, 2017; Chapter 8; pp. 145–160. ISBN 9780128094518. [CrossRef]
4. Hansen, A.D.; Iov, F.; Blaabjerg, F.; Hansen, L.H. Review of Contemporary Wind Turbine Concepts and Their Market Penetration. *Wind Eng.* **2004**, *28*, 247–263. [CrossRef]
5. Muller, S.; Deicke, M.; de Doncker, R.W. Doubly fed induction generator systems for wind turbines. *IEEE Ind. Appl. Mag.* **2002**, *8*, 26–33. [CrossRef]
6. Blaabjerg, F.; Liserre, M.; Ma, K. Power electronics converters for wind turbine systems. In Proceedings of the 2011 IEEE Energy Conversion Congress and Exposition, Phoenix, AZ, USA, 17–22 September 2011; pp. 281–290. [CrossRef]
7. Dhar, M.K.; Thasfiquzzaman, M.; Dhar, R.K.; Ahmed, M.T.; Mohsin, A.A. Study on pitch angle control of a variable speed wind turbine using different control strategies. In Proceedings of the 2017 IEEE International Conference on Power, Control, Signals and Instrumentation Engineering (ICPCSI), Chennai, India, 21–22 September 2017; pp. 285–290. [CrossRef]
8. Bebars, A.D.; Eladl, A.A.; Abdulsalam, G.M.; Badran, E.A. Internal electrical fault detection techniques in DFIG-based wind turbines: A review. *Prot. Control Mod. Power Syst.* **2022**, *7*, 18. [CrossRef]
9. Jaen-Cuellar, A.Y.; Elvira-Ortiz, D.A.; Osornio-Rios, R.A.; Antonino-Daviu, J.A. Advances in Fault Condition Monitoring for Solar Photovoltaic and Wind Turbine Energy Generation: A Review. *Energies* **2022**, *15*, 5404. [CrossRef]
10. Fischer, K.; Pelka, K.; Bartschat, A.; Tegtmeier, B.; Coronado, D.; Broer, C.; Wenske, J. Reliability of Power Converters in Wind Turbines: Exploratory Analysis of Failure and Operating Data From a Worldwide Turbine Fleet. *IEEE Trans. Power Electron.* **2019**, *34*, 6332–6344. [CrossRef]
11. Tavner, P.; Ran, L.; Penman, J.; Sedding, H. *Condition Monitoring of Rotating Electrical Machines*; Bibliovault OAI Repository, the University of Chicago Press: Chicago, IL, USA, 2008. [CrossRef]
12. Byon, E.; Ntaimo, L.; Singh, C.; Ding, Y. Wind energy facility reliability and maintenance. In *Handbook of Wind Power System*; Energy Systems; Springer: Berlin/Heidelberg, Germany, 2013; pp. 639–672. [CrossRef]
13. Shafiee, M.; Dinmohammadi, F. An FMEA-Based Risk Assessment Approach for Wind Turbine Systems: A Comparative Study of Onshore and Offshore. *Energies* **2014**, *7*, 619–642. [CrossRef]
14. Gerber, T.; Martin, N.; Mailhes, C. Time-Frequency Tracking of Spectral Structures Estimated by a Data-Driven Method. *IEEE Trans. Ind. Electron.* **2015**, *62*, 6616–6626. [CrossRef]
15. Yin, S.; Guang, W.; Karimi, H.R. Data-driven design of robust fault detection system for wind turbines. *Mechatronics* **2014**, *24*, 298–306. [CrossRef]
16. Alizadeh, E.; Meskin, N.; Khorasani, K. A negative selection immune system inspired methodology for fault diagnosis of wind turbines. *IEEE Trans. Cybern.* **2016**, *47*, 3799–3813. [CrossRef]
17. Li, M.; Yu, D.; Chen, Z.; Xiahou, K.; Ji, T.; Wu, Q.H. A Data-Driven Residual-Based Method for Fault Diagnosis and Isolation in Wind Turbines. *IEEE Trans. Sustain. Energy* **2019**, *10*, 895–904. [CrossRef]
18. Gelernter, D. *Mirror Worlds: Or the Day Software Puts the Universe in a Shoebox ... How It Will Happen and What It Will Mean*; Oxford University Press: Oxford, UK, 1993.
19. Mishra, M.; Leturiondo, U.; Salgado, O.; Galar, D. Hybrid modelling for failure diagnosis and prognosis in the transport sector. Acquired data and synthetic data. *Dyna* **2015**, *90*, 139–145. [CrossRef] [PubMed]
20. Klein, P.; Bergmann, R. Generation of Complex Data for AI-based Predictive Maintenance Research with a Physical Factory Model. In Proceedings of the 16th International Conference on Informatics in Control, Automation and Robotics—Volume 1: ICINCO, Prague, Czech Republic, 29–31 July 2019; pp. 40–50, ISBN 978-989-758-380-3. [CrossRef]

21. Leturiondo, U.; Oscar, S.; Galar, D. Validation of a physics-based model of a rotating machine for synthetic data generation in hybrid diagnosis. *Struct. Health Monit.* **2017**, *16*, 458–470. [CrossRef]
22. Liu, J.; Qu, F.; Hong, X.; Zhang, H. A Small-Sample Wind Turbine Fault Detection Method With Synthetic Fault Data Using Generative Adversarial Nets. *IEEE Trans. Ind. Inform.* **2019**, *15*, 3877–3888. [CrossRef]
23. Surrogate Optimization Algorithm—MATLAB & Simulink. Available online: https://es.mathworks.com/help/gads/surrogate-optimization-algorithm.html (accessed on 4 October 2022).

 energies

Article

HVDC Fault Detection and Classification with Artificial Neural Network Based on ACO-DWT Method

Raad Salih Jawad * and Hafedh Abid

Laboratory of Sciences and Techniques of Automatic Control & Computer Engineering (Lab-STA) Sfax, National School of Engineering of Sfax, University of Sfax, Sfax 3029, Tunisia
* Correspondence: raad.saleh@gmail.com; Tel.: +964-771-338-1461

Abstract: Unlike the more prevalent alternating current transmission systems, the high voltage direct current (HVDC) electric power transmission system transmits electric power using direct current. In order to investigate the precise remedy for fault detection of HVDC, this research proposes a method for the HVDC fault diagnostic methodologies with their limits and feature selection-based probabilistic generative model. The main contribution of this study is using the wavelet transform based on ant colony optimization and ANN to detect the different types of faults in HVDC transmission lines. In the proposed method, ANN uses optimum features obtained from the voltage, current, and their derivative signals. These features cannot be accurate to use in ANN because they cannot give reliable accuracy results. For this reason, first, the wavelet transform applies to the fault and non-fault signals to remove the noise. Then the ACO reduces unimportant features from the feature vector. Finally, the optimum features are used in the training of ANN as faulty and non-faulty signals. The multi-layer perceptron used in the suggested method consists of many layers, enabling the creation of a probability reconstruction over the inputs by the model. A supervised learning method is used to train each layer based on the selected features obtained from the ant colony optimization-discrete wavelet transform metaheuristic method. The artificial neural network technique is used to fine-tune the model to reduce the difference between true and anticipated classes' error. The input signal and sampling frequencies are changed to examine the suggested strategy's effectiveness. The obtained results demonstrate that the suggested fault detection and classification model can accurately diagnose HVDC faults. A comparison of the Support vector machine, Decision Tree, K-nearest neighbor algorithm (K-NN), and Ensemble classifier Machine techniques is made to verify the suggested method's unquestionably higher performance.

Keywords: HVDC fault detection; artificial neural network; ACO-DWT; optimization method

Citation: Jawad, R.S.; Abid, H. HVDC Fault Detection and Classification with Artificial Neural Network Based on ACO-DWT Method. *Energies* **2023**, *16*, 1064. https://doi.org/10.3390/en16031064

Academic Editors: Guang Wang, Jiale Xie and Shunli Wang

Received: 27 November 2022
Revised: 10 January 2023
Accepted: 14 January 2023
Published: 18 January 2023

1. Introduction

Due to its lower cost across long distances and capacity to transmit more power, high voltage direct current (HVDC) transmission systems have been widely used for power transmission projects with overhead transmission lines, bulk power, and asynchronous connections. The length of the lines, the surroundings of the transmission lines, and unfavorable weather conditions have all contributed to an increased error rate in HVDC transmission lines [1]. For HVDC transmission lines, current differential protection and direct current (DC) voltage reduction are commonly utilized as backup protection in addition to primary protection systems based on voltage derivatives and traveling waves. Protections based on traveling waves and voltage derivatives are vulnerable to fault resistance because they rely on the pace of the voltage change to identify problems. They frequently misdiagnose high impedance failures [2,3].

When determining a specific fault class, the neural network approaches are growing in popularity among fault prognosis techniques [4,5]. These algorithms require fault features derived from the line data (current and voltage). Even though the fault information was generated from transmission line current or voltage waveforms, a fault class cannot be determined only based on the raw signal data. Then, it was investigated how to use signal processing methods such the wavelet transforms [6], S-transform [7], or Hilbert-Huang transform [8] to separate the important properties that control the behavior of the line faults from the transmission line waveforms. As a result, increasing the neural network-based defect detection and classification model's accuracy has become one of the key study areas.

In [9], a decision tree (DT) based fault detection and classification technique for the microgrid was introduced. The discrete Fourier transform was used to extract the information. To identify the HVDC faults, DT and wavelet transformations were coupled [10]. In grid-tied DG systems, the wavelet transform (WT) and the S-transform can also be used to detect disturbances [11]. To identify the different types of faults, a learning model that combines the naive classifier, support vector machine (SVM), and Extreme Learning Machine (ELM) based on the traits returned by the Hilbert–Huang transform has been utilized [12]. ELM and discrete wavelet transform (DWT) were combined in [13] to detect, classify, and identify a microgrid's sections. For the microgrid's flaw identification and classification, Reference [14] presented a semi-supervised model; first demonstrated in [15], a Taguchi-based artificial neural network (ANN) using DWT. The shallow design constrains these neural networks- and machine learning-based techniques. They employ the ability of the complicated non-linear properties of the HVDC to learn. These methods cannot combine the advantages of numerous aspects with perfection since there are no hidden layers.

Raad Salih et al. [1] used the gray wolf optimization method based on ANN to detect the fault in the HVDC system. They used the gray wolf optimization algorithm to select the best features extracted from the voltage and current signals.

The metaheuristic methods are used in [16,17], and deep learning methods are implemented in [18,19]. The improved power quality and fault detection are presented in [20]. The economic dispatch in the HVDC system is presented in [21]. The protection in sensitive load is investigated in [22,23].

In all approaches mentioned in the literature, feature selection has not been used. Moreover, the wavelet transform cannot give accurate results because some features cannot be reliable for fault detection. It is a significant disadvantage of the previously used methods in the literature. In this study, by combining the WT-ACO, the problem is solved.

This research offers an ANN with numerous layers of hidden units to address the issues and provide a method for learning the intricate non-linear feature of the HVDC in order to increase classification accuracy. ANN was initially used to identify aircraft engine problems. After that, research into diagnosing faults in gearboxes, rolling bearings, and reciprocating compressor valves expanded quickly [24–27]. ANN is a stack of ACO-DWT that deepens the network and makes it possible for the model to extract features in an adaptive manner. ANN can work with non-linear data [28–30]; therefore, it can more precisely classify faults in the microgrid domain.

The proposed network performs fault diagnostics on the HVDC system with phase-to-phase and phase-to-ground fault breaker systems using voltage and current waveform data as input. The features are extracted from the raw signal samples using a discrete wavelet transform tool. This research also suggests an extension of ant colony optimization to select the best features created by DWT with the dropout approach to improve the accuracy performance of fault detection. The dropout approach considerably improves the fault detection accuracy performance against a traditional method. The following are the article's primary conclusions:

- Based on ANN and the DWT [31,32], we develop a model for fault detection and classification of HVDC that effectively extracts the pertinent short circuit fault attribute from the faulted line signals;
- Using a supervised pre-trained method, the trained model was created with multi-layers of the network that prevents overfitting of the training data;

- To evaluate the efficacy of the suggested approach, we examine both islanded and grid-connected/tied operating modes as well as the two HVDC (radial and loop) typologies;
- The dropout method is incorporated with the suggested network to ensure that the generated ANN model performs well in noisy environments.

The protection block begins with the fault detection unit as its first component. Therefore, a quick and dependable solution is required to detect flaws in HVDC protection systems. In this study, an ANN-based on Ant Colony Optimization and wavelet transform is employed to react quickly to detect faults. ANN algorithm requires some time to learn, but after it has completed the learning step, the trained network can move on to the fault detection stage. ANNs may detect faults much faster during the testing phase than traditional logic techniques. The importance of developing fault detection techniques is that they increase accuracy, sensitivity, and reliability, and that is what the transmission authority needs to decrease time and cost for finding and repairing the faults in the HVDC system.

The order of the paper is as follows. The material and method are presented in Section 2. The design of the suggested approach for categorizing HVDC faults and the necessary materials, is also described in Section 2. The performance analysis of the suggested system is shown in Section 3. Section 4 will finalize this paper with a conclusion and future work.

2. Material and Method

Shunt faults and series faults are the two basic categories under which HVDC power line faults fall [33]. A series fault, also known as a simple break in one or more conductors, occurs when there is an imbalance in the series impedance on the line. Power transmission from one location to another is not directly related to this kind of failure. In contrast, the three-phase power network regularly experiences shunt faults during power transmission, which are subsequently categorized as phase-to-phase (PP), phase-to-ground (PG), and two PG faults (2PG).

A single line-to-ground fault can occur on any phase line of a three-phase power line if it meets the neutral line or hits the ground. The problem brought on by strong winds or trees falling on power lines is also known as a short circuit fault [34]. Figure 1A–C depicts three HVDC system stages and shows three forms of single line-to-ground faults [34].

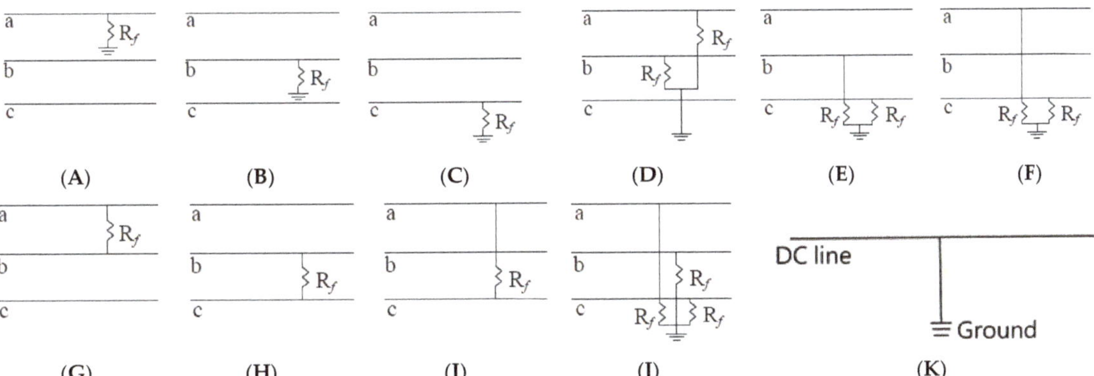

Figure 1. HVDC system fault classes (**A**) a-g, (**B**) b-g, (**C**) c-g, (**D**) ab-g, (**E**) bc-g, (**F**) ac-g, (**G**) a-b, (**H**) b-c, (**I**) a-c, (**J**) a-b-c-g, (**K**) DC fault [34].

A two-PG fault occurs when two lines of a power line fall to the ground. This fault has more asymmetry and a higher fault current amplitude than the line-to-line fault. If this issue is not fixed right away, it could develop into a three-line to-the-ground fault, which is much more dangerous than other fault types. The ab-g, b-g, and a-g faults are depicted in Figure 1D–F, where Rf is the fault resistance [34].

A short circuit between any two lines in a three-phase system causes this kind of failure. One of the significant aspects of this asymmetrical fault is the difficulty in predicting the upper and lower boundaries of the fault impedance due to its magnitude varying over a large range. Three distinct PP fault types are displayed in Figure 1G–I. in contrast, three phases to ground fault appear in Figure 1G. Figure 1K illustrates the DC line to the ground fault. This paper studies the DC faults that may happen in the HVDC system's DC line and AC faults that occur in the AC side of one of two terminal LCC HVDC systems under study.

2.1. System Modelling

This research proposes a probabilistic, generative network-based framework for detecting and classifying HVDC system faults. The sound or healthy condition was utilized to create a type of fault that provided 11 fault types for the unhealthy or fault detection plan. This sort of fault encompassed all short circuit fault scenarios and the good state of the phase. It was assumed that the classifier's nature would be sound or error-free under typical conditions. A bad or fault event was detected when the classifier output was changed to a particular fault class. Extraction of the fault features from the raw signals was required for the suggested method's training. Each fault signal's energy was unique and determined by the system parameters, such as the fault distance and resistance. The DWT was used to independently assess the variation in each phase's raw signals. After that, each signal's energy was estimated in order to create the necessary dataset. Figure 2 shows the HVDC system under test, and Table 1 includes the parameters of the system Voltage Source Converters (VSC) that are used in modern HVDC; nevertheless, the model used in this study makes use of thyristors. The literature contains well-known and cutting-edge protection techniques for thyristor-based two-terminal HVDC systems [35–37]. VSC-based systems, particularly multi-terminal DC systems, are currently facing protection issues. The GWO approach, the study's main topic, is used to assess the features and choose the appropriate voltage and current signal format.

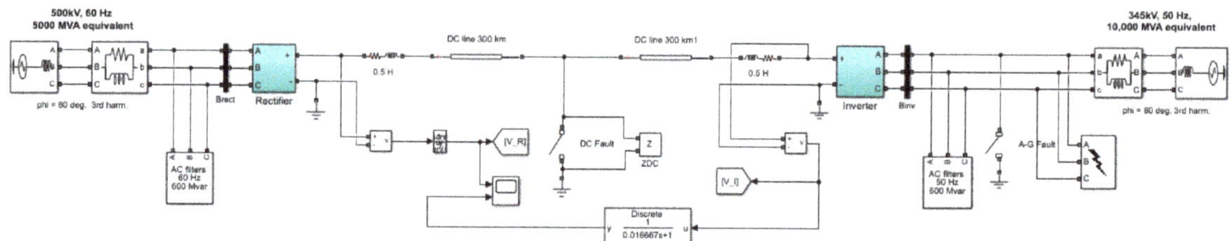

Figure 2. LCC-HVDC system under study.

Table 1. Parameters of the HVDC system under study.

Parameter	Value
AC system one voltage	500 kV
AC system two voltage	345 kV
AC system one frequency	60 Hz
AC system two frequency	50 Hz
DC line length	300 km
DC line voltage	500 kV
DC power transmitted	1000 MVA

Table 1. *Cont.*

Parameter	Value
DC current	2000 A
Smoothing reactor at the rectifier	0.5 H
Smoothing reactor at the inverter	0.5 H

2.2. *Effect of Fault Distance on Signal Energy*

A defective occurrence could happen anywhere along the HVDC systems. The proposed ANN's training method, which varied the network, could inspect the signal across the entire HVDC system thanks to the fault distance. In order to create the sample data, the fault event's position was changed from the current and voltage waveforms, and the distance measurements ranged from 1 to 19 km with a 0.5 increment. The change in the original signal produced different signal energy lengths, which ultimately represented different features.

2.3. *Wavelet Transform for the Generation of Fault Feature*

In this work, the qualities of a particular section of a signal during a quick change in the signal were examined using the DWT. The WT works by breaking down a signal into a sequence of temporal components. A time series waveform receives significant guidance from the time series faulty sections, which shield a specific frequency range. Here is an example of how the WT approach represents, during a fault event, the fault attribute of a particular piece of the faulty signal. a group of low-pass (LPF) and high-pass (HPF) filters are used in DWT to process signals (LPF). The signal is broken up into detail (Det) and approximation (App) coefficients as a result of the LPF's analysis of the examination of the high-frequency domain signal by the HPF and the low-frequency domain signal. The fault signal's large- and small-scale frequency components are represented utilizing the App coefficient. Therefore, the fault signal is represented by its small- and large-scale frequency components by the Det coefficient. The subsequent App replicates this decomposition procedure, dividing a fault signal into several lesser-resolution pieces.

2.4. *Model for Proposed Hierarchical Generative Faults*

This section introduces the ANN framework for HVDC fault detection and classification. An arrangement of constrained machine learning makes up the proposed ACO-DWT. Using the ACO-DWT, the features are extracted and selected accurately.

2.5. *Wavelet Analysis*

A wavelet transform function is the display of lower-frequency signals larger and high-frequency signals narrower when wavelet detection is present. Despite similarities, there are a few significant differences between the Fourier transform and the wavelet transform. The signal is divided into sines and cosines via the Fourier transform. On the other hand, the wavelet transform can be used with elements in both Fourier and real spaces. The temporal widths of the wavelet transform can be changed to match the frequency. This attribute of frequency width auto-tune is most helpful when assessing electromagnetic transients that have superimposed on the frequency power components are high-frequency components [38]. Typically, the wavelet transform looks like this:

$$WT(f,a,b) = \frac{1}{\sqrt{2}} \int_{-\infty}^{\infty} f(x)\psi^*\left(\frac{t-b}{a}\right) \quad (1)$$

where a and b are the function constants, which are also known as the scaling and translation parameters, and (*) is the complex conjugate of the wavelet function ψ. Continuous wavelet transforms (CWT), and discrete wavelet transform are the two subcategories of the wavelet transform (DWT). The wavelet transform is derived into the correlated wavelet transform (CWT), which uses redundant wavelets and arbitrary scales. By breaking down the signal into orthogonal sets using a discrete set of wavelet scales, the discrete wavelet transform (DWT) is produced. The discrete wavelet transform is obtained using the following expression (DWT).

$$DWT(f, m, m) = \frac{1}{\sqrt{a}} \sum_{k} f(k) \psi^{*} \left(\frac{n - k a_0^m}{a_0^m} \right) \tag{2}$$

The parameters "a" and "b" are swapped out for a_0^m and ka_0^m, where k and m are integers compared to the term 2.17. The DWT functions as a bank of low-pass and high-pass filters that provide low-pass and high-pass subbands for the signal. The low-pass subband is subjected to the same procedure to create narrower low-pass and high-pass sub-bands. Wavelet transforms, either continuous or discrete, can be used to assess the estimated distance to the fault. In order to define a mother wavelet from a voltage transient waveform, a continuous wavelet transform is utilized in the research of fault location in power networks [39]. However, the analysis of this study can give good results with just the discrete wavelet transform (DWT). Figure 3 shows the wavelet family featuring the daubechies3 (db3) wavelet mother, which is utilized to decompose voltage waveforms registered by DFRs into its five coefficients (WTCs). Due to their abundance of high-frequency content, the WTCs at level 1 (D1) are subsequently analyzed to identify the times of arrival (ToAs). These signals are finally squared to create WTC2, as was conducted in [38], to reduce noise in WTCs.

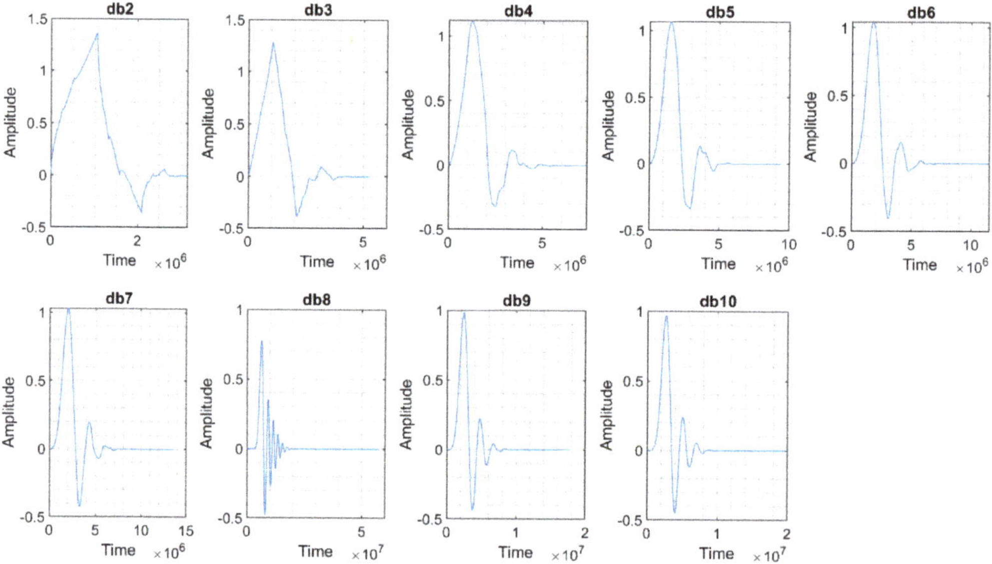

Figure 3. Wavelet Daubechies' family.

The signals with and without faults were constructed to investigate the HVDC fault detection in this work. Multiple signals with various AC and DC fault types were devised for this purpose. These signals are used to extract the 12 properties that depend on the voltage, current, and their individual components. Some of these features are inappropriate for ANN training [18,19], and employing them will lead to mistakes and lower detection accuracy. The best and most precise characteristics ought to be chosen for this purpose. Thus, the feature selection uses the ACO approach, which was first introduced in [40]. Figure 4 depicts a summary of the suggested technique.

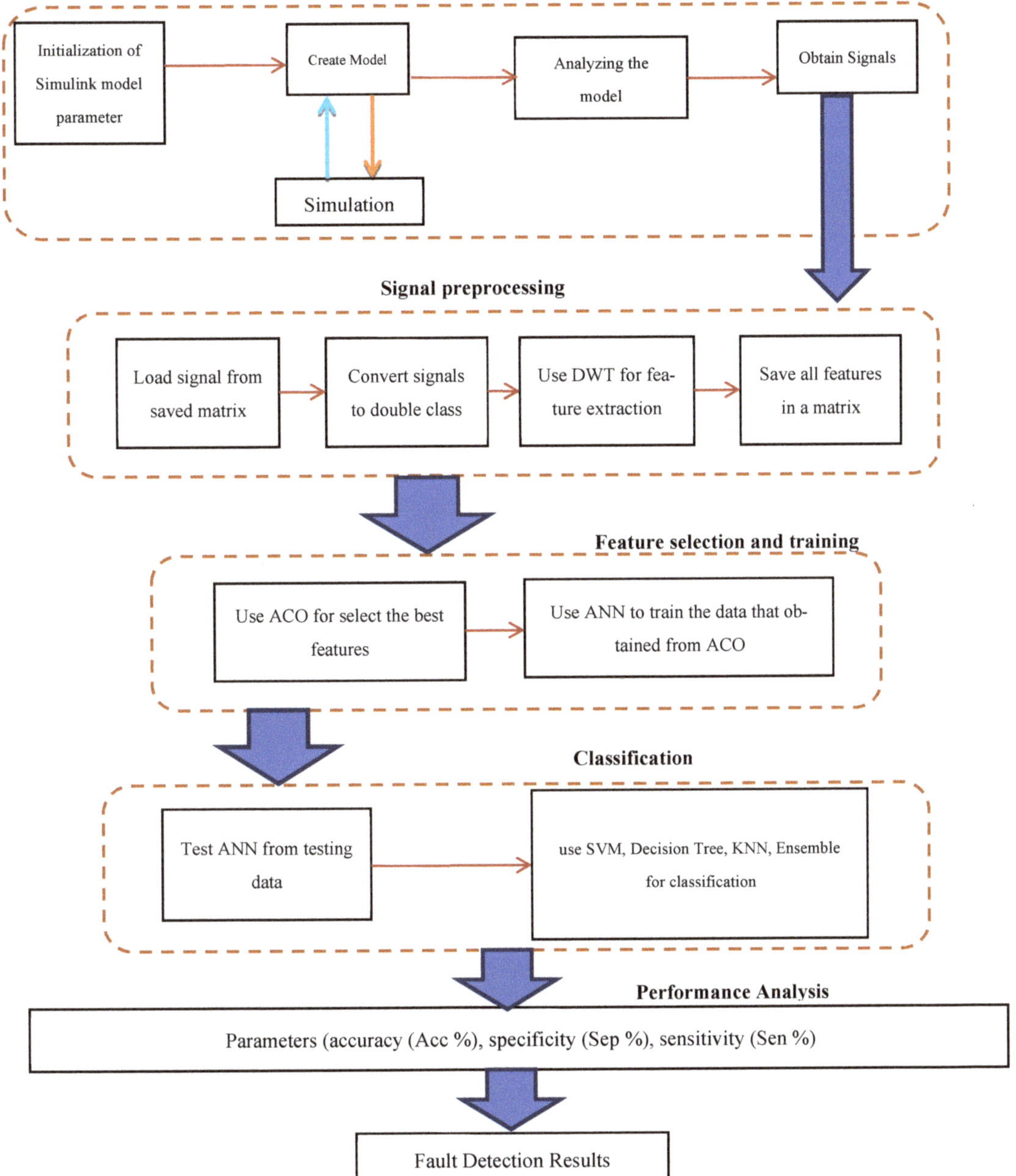

Figure 4. Summary of the suggested technique for finding HVDC faults.

As shown in Figure 4, Simulink-MATLAB manually generates the 11 faulted signals, including the AC and DC faulted signals. Next, the output signals are obtained for each fault. The neural network may occasionally make errors in fault identification because these signals contain many characteristics, the majority of which are unsuitable for training. Therefore, the best and most practical characteristics must be chosen to train the network.

The best features are selected using the ACO approach, and the neural network is trained using these features. On the right side of the flowchart is the ACO method scenario.

The main objective of this study is to use the optimum features of the fault and non-fault signals by using the Wavelet transform based on the ACO algorithm. The wavelet transform removes the noisy signal from the current and voltage signals obtained from the system. Furthermore, ACO uses to find the optimum features that affect the training of ANN to recognize the fault and non-fault signals.

First, a visual representation of every feature in the S dataset is presented. All nodes are connected to one another and are referred to as nodes. The number of ants and the number of repetitions should then be determined [41]. The value is known as the pheromone trail, and all of its values are initially set to a fixed value of one at the beginning of the algorithm. also known as the value of heuristic information, is equal to the reciprocal distance between the qualities [42], which will be determined in this article using the two approaches, FC and FF.

The algorithm is usable after establishing the initial values. The ant is initially placed on a node at random in each iteration. The rule of transfer is applied to derive the following ninety, as indicated in Equations (3) and (4):

$$P_i^k(t) = \frac{|\tau_i(t)|^\alpha * \times |\eta_i(t)|^\beta}{\sum_{u\in j^k} |\tau_i(t)|^\alpha \times |\eta_i(t)|^\beta} \quad if(q > q_0) \tag{3}$$

$$j = \max_{u\in j^k}(\tau_i(i)^\alpha \times \eta_i(i)^\beta) \qquad if(q < q_0) \tag{4}$$

The α and β values are chosen in order to increase the efficacy of τ and η. The ant has not yet encountered the attributes in j^k, and the only trait they have is zero. The parameter q_0, whose value is a random number between 0 and 1, is crucial in selecting whether to use the greedy or probabilistic approach.

The amount of pheromone collected from the scan should be updated in accordance with Equation (5) when the n ant has finished the node scan:

$$\tau_i(t+1) = (1-\rho)\tau_i(t) + \sum_{i=1}^n \Delta\tau_i^k(t) \tag{5}$$

To lessen the effect, the value of the average number of nodes chosen for the Filter technique is equal to $\Delta\tau_i^k$ is determined, which is the reverse of the error achieved using the Wrapper technique [41,42].

2.6. Criteria for Distance or Similarity of Features

Two types of relationships exist between two random variables: linear and non-linear. The correlation coefficient formula is the most well-known formula for calculating linear variables. To compute non-linear variables, they employ information theory and the entropy approach. The correlation coefficient technique has the drawback of being ineffective with batch and non-numerical data, but the entropy method performs well [43].

A discrete or continuous random variable's uncertainty is measured using entropy or irregularity criteria. The discrete random variable $X = (x_1, x_2, \ldots, x_n)$ has an entropy of $H(X)$ that is determined using Equation (6).

$$H(X) = -\sum_{i=1}^n p(x_i)\log(p(x_i)) \tag{6}$$

where $p(x_i)$ is the probability value of x_i happening on the entire set.

In accordance with Equation (7), the two discrete random variables' entropies should be calculated X and Y.

$$H(X,Y) = -\sum_{i=1}^n \sum_{j=1}^n p(x_i, y_i)\log(p(x_i, y_i)) \tag{7}$$

The conditional entropy of X to condition Y is determined using Equation (8).

$$H(X|Y) = -\sum_{i=1}^{n}\sum_{j=1}^{n} p(x_i, y_i)\log(p(x_i|y_i)) \quad (8)$$

The aforementioned formulas' goal is to determine the information factor (IF). The IF criterion, which is in agreement with Equation (9), is utilized to analyze how dependent the two variables are:

$$I(X,Y) = H(X) - H(X|Y) \quad (9)$$

The two variables are independent if the value of IF is zero, and the larger this value, the more dependent X and Y are [44]. The correlation between the information coefficient and entropy is depicted in Figure 5.

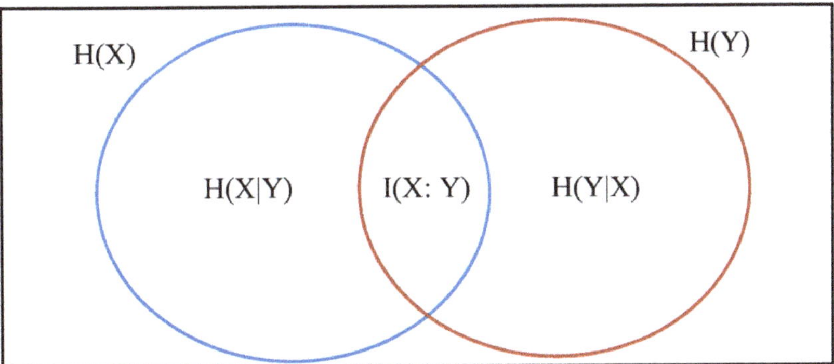

Figure 5. Relationship between information coefficient and entropy.

The symmetrical uncertainty (SU), or normalized form of IG, used in this work is compatible with Equation (10). This formula's benefit is the normalcy of the two variables' dependence between 0 and 1. The two variables are dependent if the value of SU is close to one and independent if the value of SU is close to zero.

$$SU(X,Y) = \frac{2 * I(X,Y)}{H(X) + H(Y)} \quad (10)$$

Two criteria SU_{FC} and SU_{FF} are employed in this paper to calculate the η [43].

The reliance of each attribute on the class is the definition of the SU_{FC} criterion. The more vital and desirable that characteristic will be, the closer this number is to one.

$$\eta_i = \frac{1}{1 - SU_{FC}} \quad (11)$$

The term SU_{FF} refers to the interdependence of two qualities. If its value is very near to one, it indicates that the two traits are quite comparable, and we might consider eliminating one of the features.

$$\eta_i = \frac{1}{SU_{FF}} \quad (12)$$

When choosing attributes, we try to keep class-related attributes and remove redundant or unnecessary attributes. The objective is to select features that have SU_{FC} higher and SU_{FF} lower values [45].

In the first step of this research study, the voltage and current signals are generated. Then these signals are analyzed to determine the characteristics of the voltage and current signals. The ACO-DWT approach is utilized to choose the best and most effective characteristics. In order to identify the most useful features with which to train ANN, the proposed method based on the ACO algorithm is used.

3. Results and Discussion

This section uses various parameter adjustments to show how well the suggested fault detection and classification technique performs. Different magnitudes were disclosed by the fault current and voltage signals in grid-connected and islanded modes. As a result, creating a uniform fault classification scheme was challenging. As a result, the effectiveness of the suggested strategy was examined individually under different operating modes and system topologies. Three factors were taken into consideration when determining the accuracy:

(a) How well the system performed, whether only using the current waveform, voltage waveform, or both waveforms together, depending on the type of input signal;
(b) The sampling resolution, or how well the system performed when using different data acquisition rates;
(c) The fault signal with noise present, or how well the system performed when noise was present in the sampled signal. The suggested and existing FDC techniques' accuracy were compared to show that the suggested classifier has superior short circuit fault classification capabilities.

In machine learning, a list of data samples is used to test the model's performance and should be different from the training data in order to determine how effective a learning model is. The current and voltage waveforms for each dataset were divided into 1716 samples, which were then combined and shuffled before being randomly selected to assess the effectiveness of the suggested strategy. When the 11 distinct fault classifiers were entered into the 11×11 matrix's x- and y-axis, the confusion matrix (CM) was used to simulate the performance of ANN for Lines 1–3 under various system configurations and HVDC operating modes. The vertical levels represent the projected fault class, while the horizontal levels indicate the actual class. The true positive (TP), true negative (TN), false positive (FP), and false negative (FN) counts are also reported in the confusion matrix and are defined as:

TP: The classifier correctly predicted a label and is a member of the original class;

TN: The classifier successfully predicts a label even when it does not fall under the initial category;

FP: The classifier predicts a label to be positive even when it does not belong to the original category;

FN: The label that the classifier predicts would be negative but belongs to the original class.

First and foremost, according to the CM, most of each system configuration's fault classes were assigned correctly. The classification accuracy of the average confusion matrix (Acc) was used as the first accuracy measurement criterion Equation (13).

$$Accuracy = \frac{(TP + TN)}{(TP + FP + TN + FN)} = \frac{N_{TD}}{N_{CC}} \tag{13}$$

Here, N_{TD} displays the total number of data used to create the model. N_{CC} denotes the number of correctly categorized data. The remainder of the HVDC system might function similarly to the proposed work.

According to the findings, the proposed classifier's grid-connected radial mode operation had the best accuracy, which was 99.70%. The classifier performed better than 99.5% for the other system settings, which was as expected. The Confusion matrix for different classifiers, SVM, Decision Tree, K-NN, and Ensemble method, is shown in Figure 6.

The average accuracy for various system setups is shown in Figure 7.

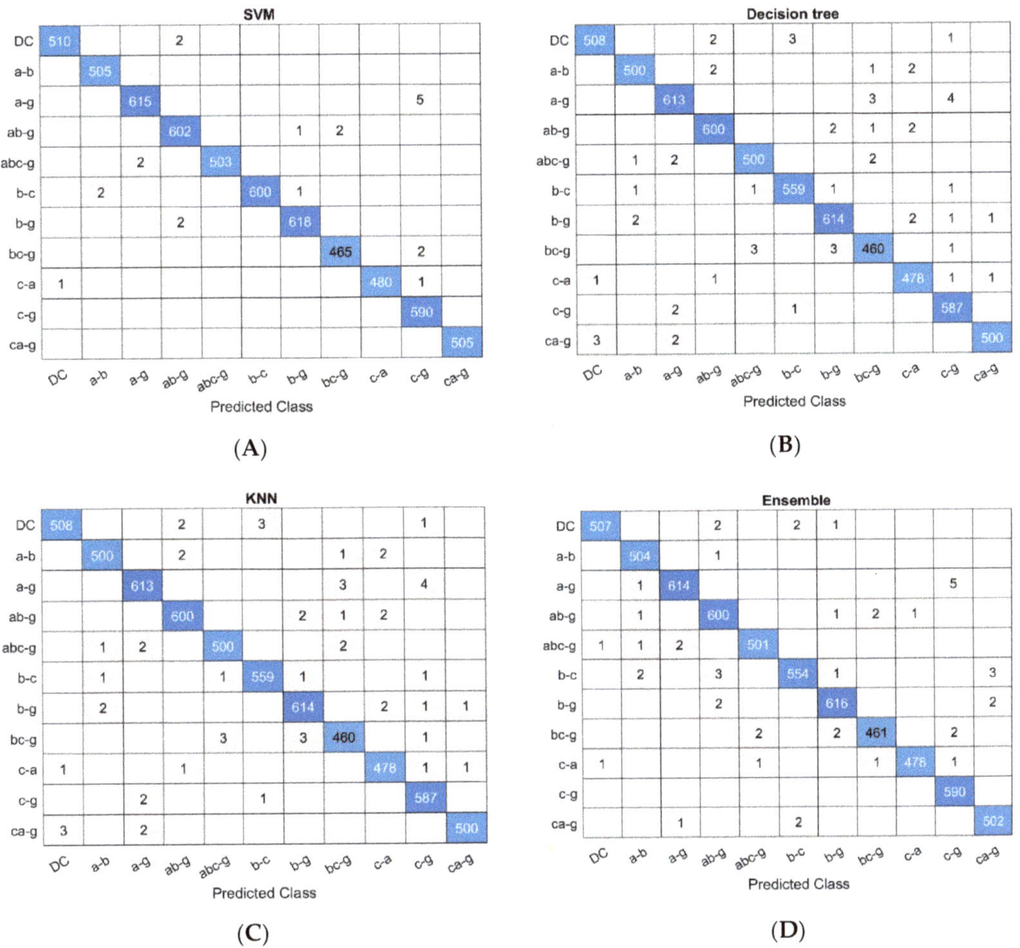

Figure 6. Confusion matrix for different classifiers (**A**) SVM, (**B**) Decision Tree, (**C**) K-NN, and (**D**) Ensemble.

The average accuracy, however, was unable to provide a complete analysis of the model's performance. The sensitivity, specificity, and accuracy were then used to assess the classification performance in order to determine how the classifier handled different fault types. The mentioned criteria are regarded as ideal when it equals one and worst when it equals zero. The sensitivity, also referred to as the positive predictive value, is defined as follows:

$$Sensitivity = \frac{TP}{(TP + FN)} \tag{14}$$

A good classifier should have a precision value of one. From Equation (14), the precision value declines as the FP rises, which is unexpected for a strong classifier. Specificity, another statistic that is also referred to as the true positive rate or the classifier's sensitivity, is defined as follows

$$Specificity = \frac{TN}{(TN + FP)} \tag{15}$$

The best classifier's sensitivity value should be 1, just like the precision. For this metric, the recall value fell as the FN grew, which was also contrary to expectation. As a result, accuracy, which considers both true positive and true negative, was used as another performance evaluation indicator. Given that Table 2 shows the voltage and current signals, the recommended classifier's greater accuracy demonstrated that it had fewer false positives and negatives. Additionally, each fault class's categorization accuracy (user accuracy) made it clear that the classifier had a high degree of accuracy in its capacity to categorize the problems.

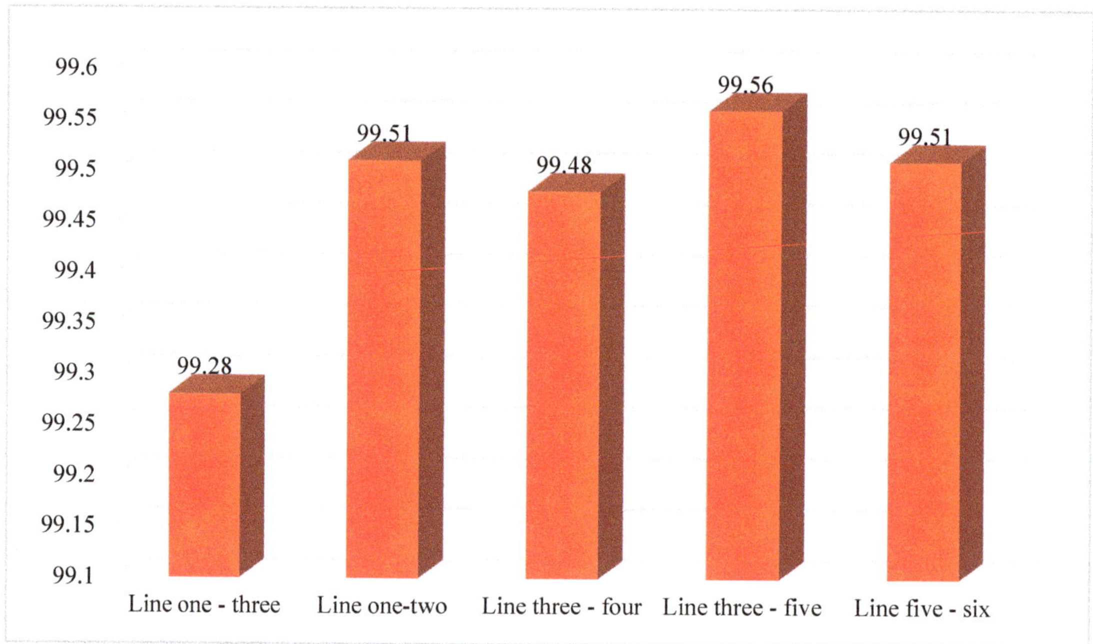

Figure 7. Average accuracy for various system setups.

Table 2. Accuracy, sensitivity, and specificity of the proposed classifier.

Fault Class	Accuracy (%)	Sensitivity (%)	Specificity (%)
a-g	98.95	98.83	99.05
b-g	98.58	98.31	98.86
c-g	99.36	98.87	99.86
ab-g	98.88	99.58	98.16
bc-g	98.60	98.40	98.80
ac-g	99.10	99.25	98.93
a-b	98.77	98.76	98.79
b-c	98.38	98.45	98.29
a-c	98.59	98.42	98.74
abc-g	99.48	98.95	100.00
DC	98.91	99.56	98.29

Figure 8 is a graphical depiction of these findings.

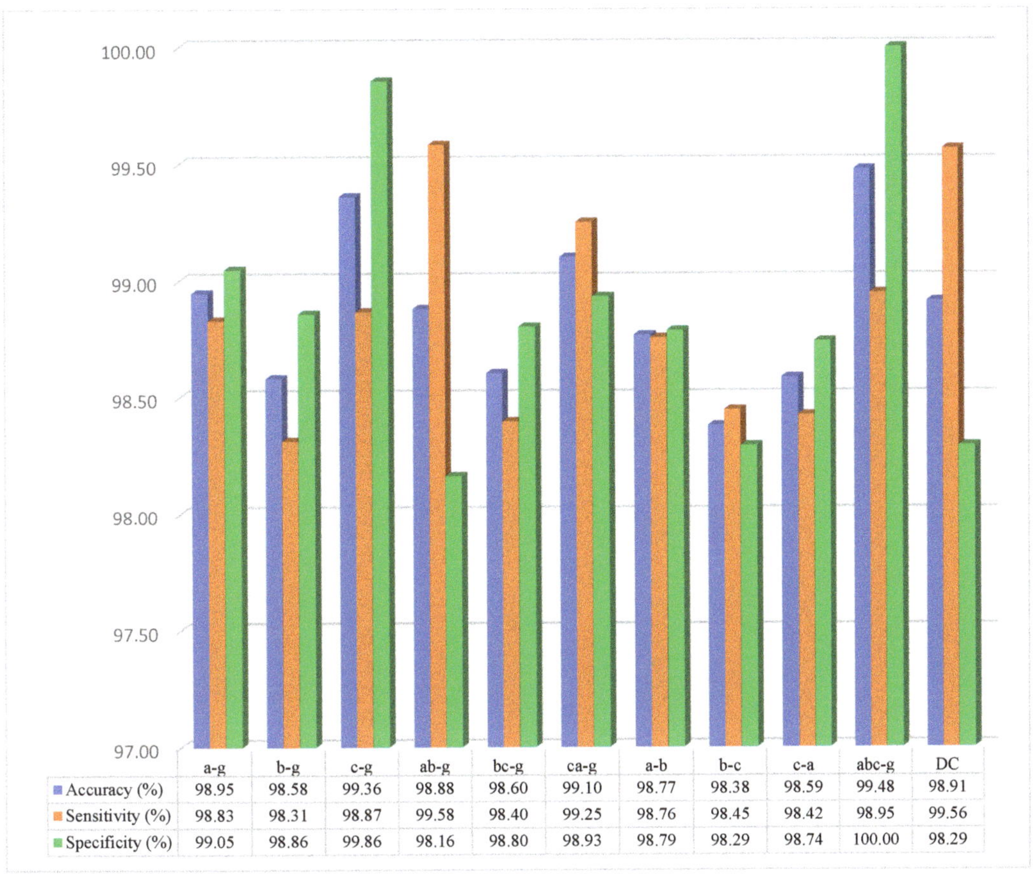

	a-g	b-g	c-g	ab-g	bc-g	ca-g	a-b	b-c	c-a	abc-g	DC
Accuracy (%)	98.95	98.58	99.36	98.88	98.60	99.10	98.77	98.38	98.59	99.48	98.91
Sensitivity (%)	98.83	98.31	98.87	99.58	98.40	99.25	98.76	98.45	98.42	98.95	99.56
Specificity (%)	99.05	98.86	99.86	98.16	98.80	98.93	98.79	98.29	98.74	100.00	98.29

Figure 8. Graphical chart illustration.

Due to the lack of both signals at the same time, the system must perform the categorization tasks using voltage or current waveforms. Different input signal types and sampling rates were employed to examine the fault classification performance of the proposed classifier. The input signal types used in this inquiry were the voltage waveform, the current waveform, and the combined current and voltage waveform. The SF used were 2, 5, 10, 15, and 20 kHz. Five times through the classification process, the findings for an SF and a certain signal type were determined. To obtain the final findings presented in Figure 9, the mean value of the accuracies was calculated.

The improvement in classification accuracy was anticipated since a short circuit fault class with a greater SF carries more specific fault information. The three-phase current waveform performed better for classification at lower sample rates than the three-phase voltage waveform. This scenario was anticipated in part because, for a given fault class, compared to the current waveform, the voltage waveform contained less information about low-frequency faults.

On the other hand, the voltage waveform had a few spare incorrect transients that could be used to examine the short circuit fault's specifics more thoroughly. The aforementioned analysis suggests that relying just on current or voltage waveforms would not produce the requisite precision. Given that both the three-phase current and voltage intentions utilized specific short circuit fault information, if both waveforms were taken into account simultaneously, the designated frequency level may be obtained with a higher fault categorization performance. The study's results showed that using only the current or voltage waveforms to classify data yielded poor results; however, their fusion produced a classification accuracy of more than 99% at the high-frequency range level taken into account, validating the efficacy of the proposed FDC model. The remaining HVDC system under evaluation can show similar categorization results. Table 3 compares the accuracy, sensitivity, and specificity of the entire fault classification system for the HVDC fault classification with the suggested technique and alternative methods.

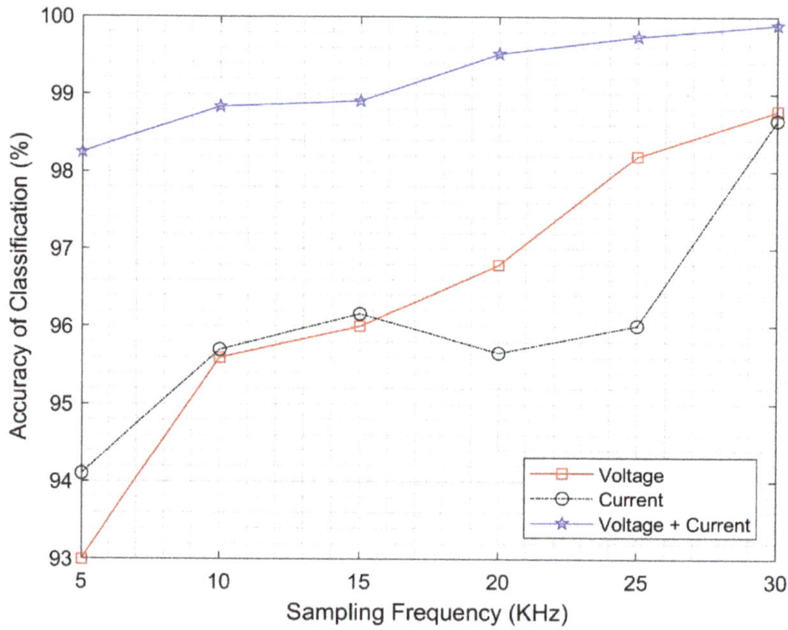

Figure 9. Accuracy of the classification for the proposed classifier.

Table 3. The proposed method's accuracy, sensitivity, and specificity.

ANN-Machine Techniques	Accuracy (%)	Sensitivity (%)	Specificity (%)	TP	TN	FP	FN
MLP	98.86	98.39	99.35	98.28	98.23	0.64	1.61
RBF	98.65	98.45	98.86	97.93	97.35	1.13	1.54
LVQ	98.78	98.17	99.39	98.57	98.49	0.60	1.83
SOM	98.30	98.23	98.38	97.88	97.56	1.61	1.76
ANN-DWT-ACO-SVM	99.49	99.05	99.94	98.16	98.28	0.06	0.94
ANN-DWT-ACO-DT	**99.60**	**99.25**	**99.96**	**99.62**	**98.14**	**0.04**	**0.75**
ANN-DWT-ACO-KNN	99.48	99.69	99.27	98.88	99.65	0.72	0.31
ANN-DWT-ACO-Ensemble	99.45	99.13	99.77	97.79	97.40	0.22	0.86

The best aspects of voltage, current, and derivatives are utilized in the suggested strategy. It is contrasted with the multi-layer perceptron (MLP), radial basis function (RBF), learning vector quantization (LVQ), and self-organizing map (SOM) neural networks. Results from the studies showed that the advised approach, ANN, RBF, LVQ, and SOM had accuracy values of 98.86, 98.65, 98.78, 98.30, 99.49, 99.60, 99.48, and 99.45, respectively. The proposed method had the highest accuracy because when the feature selection component of the ACO-DWT algorithm, which is based on the Decision tree classifier, was utilized, the accuracy rose to 99.60%.

The graphical illustration of the results is shown in Figures 10 and 11.

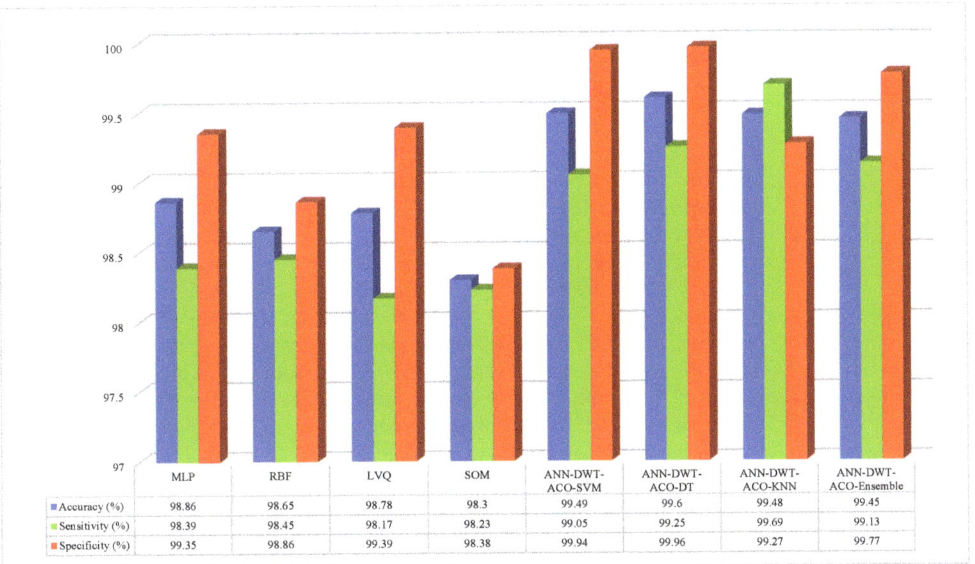

	MLP	RBF	LVQ	SOM	ANN-DWT-ACO-SVM	ANN-DWT-ACO-DT	ANN-DWT-ACO-KNN	ANN-DWT-ACO-Ensemble
Accuracy (%)	98.86	98.65	98.78	98.3	99.49	99.6	99.48	99.45
Sensitivity (%)	98.39	98.45	98.17	98.23	99.05	99.25	99.69	99.13
Specificity (%)	99.35	98.86	99.39	98.38	99.94	99.96	99.27	99.77

Figure 10. Graphical illustration of the accuracy, sensitivity, and specificity.

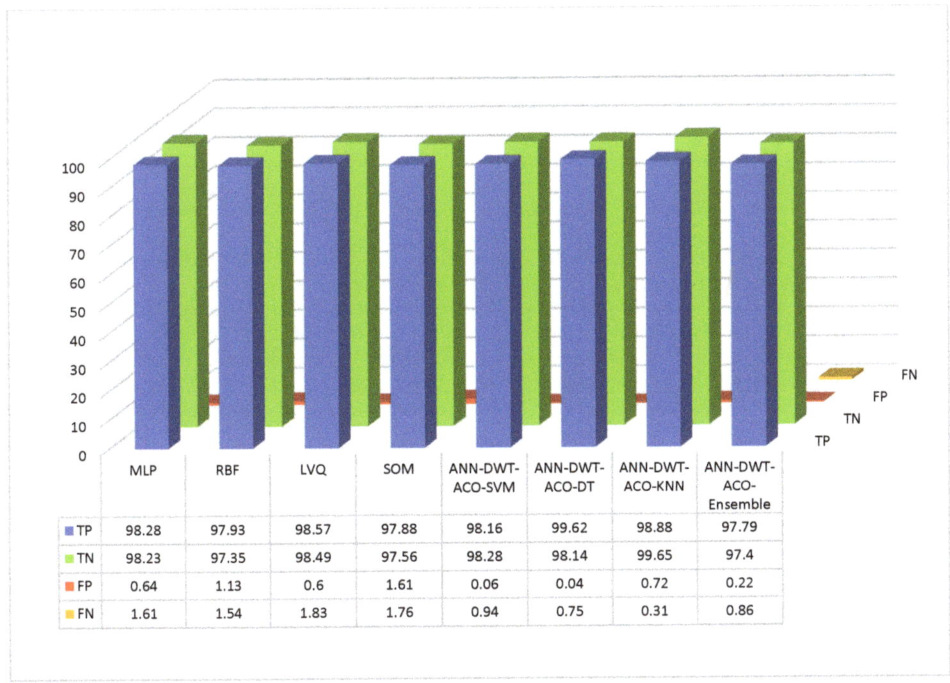

	MLP	RBF	LVQ	SOM	ANN-DWT-ACO-SVM	ANN-DWT-ACO-DT	ANN-DWT-ACO-KNN	ANN-DWT-ACO-Ensemble
TP	98.28	97.93	98.57	97.88	98.16	99.62	98.88	97.79
TN	98.23	97.35	98.49	97.56	98.28	98.14	99.65	97.4
FP	0.64	1.13	0.6	1.61	0.06	0.04	0.72	0.22
FN	1.61	1.54	1.83	1.76	0.94	0.75	0.31	0.86

Figure 11. Graphical illustration of the TP, TN, FP, and FN.

4. Conclusions

In order to identify the optimum characteristics of the voltage and current signal to utilize in ANN to train the system, this paper suggested the ACO for feature selection. For the grid-connected and island modes of the HVDC, the ACO approach is robust in identifying the best aspects of the signals to detect and classify the faults. The suggested approach ensured that the model automatically recognized and analyzed abnormal signals pertaining to various HVDC failures. This was accomplished by measuring the voltage and current waveforms separately and utilizing feature extraction to compare them to variations in the line characteristics. The proposed method's usefulness as a generalized model that worked at different sampling frequencies was confirmed by using both the current and voltage parts for fault diagnosis and classification. The suggested technique's efficiency was assessed using various experiments, such as those examining the influence of signal type. The results demonstrated that the suggested fault detection and classification model correctly recognized and categorized short circuit faults for all fault categories with an accuracy close to 99.60%. The best scenario was obtained from the ANN-DWT-ACO-DT method, with 99.60, 99.25, 99.96, 99.62, 98.14, 0.04, and 0.75 for accuracy, sensitivity, specificity, TP, TN, FP, and FN, respectively. Using both the voltage and current waveforms within the tested frequency range showed the model's impressive performance. The authors advise employing the Fourier transform for feature extraction of the current and voltage signals and various metaheuristic techniques to identify the accuracy rates for defect detection situations.

Author Contributions: Conceptualization, R.S.J. Methodology, R.S.J. Software R.S.J. Validation, H.A. Investigation R.S.J. Writing—original draft preparation, R.S.J. Writing—review and editing, H.A. Visualization, R.S.J. Supervision, H.A. All authors have read and agreed to the published version of the manuscript.

Funding: The authors declare that there was no funding for this work.

Data Availability Statement: No new data were created or analyzed in this study. Data sharing is not applicable to this article.

Conflicts of Interest: The authors declare no conflict of interest.

References

1. Jawad, R.S.; Abid, H. Fault Detection in HVDC System with Gray Wolf Optimization Algorithm Based on Artificial Neural Network. *Energies* **2022**, *15*, 7775. [CrossRef]
2. Liu, J.; Tai, N.; Fan, C.; Huang, W. Protection scheme for high-voltage direct-current transmission lines based on transient AC current. *IET Gener. Transm. Distrib.* **2015**, *9*, 2633–2643. [CrossRef]
3. Zhang, Y.; Li, Y.; Song, J.; Li, B.; Chen, X.; Zeng, L. Novel protection scheme for high-voltage direct-current transmission lines based on one-terminal transient AC voltage. *J. Eng.* **2019**, *2019*, 4480–4485. [CrossRef]
4. Yadav, A.; Dash, Y. An overview of transmission line protection by artificial neural network: Fault detection, fault classification, fault location, and fault direction discrimination. *Adv. Artif. Neural Syst.* **2014**, *2014*, 230382. [CrossRef]
5. Singh, J.; Goyal, K.K.; Kumar, R.; Gupta, V. Development of artificial intelligence-based neural network prediction model for responses of additive manufactured polylactic acid parts. *Polym. Compos.* **2022**, *43*, 5623–5639. [CrossRef]
6. Hong, Y.-Y.; Wei, Y.-H.; Chang, Y.-R.; Lee, Y.-D.; Liu, P.-W. Fault detection and location by static switches in microgrids using wavelet transform and adaptive network-based fuzzy inference system. *Energies* **2014**, *7*, 2658–2675. [CrossRef]
7. Gush, T.; Bukhari, S.B.A.; Mehmood, K.K.; Admasie, S.; Kim, J.-S.; Kim, C.-H. Intelligent fault classification and location identification method for microgrids using discrete orthonormal stockwell transform-based optimized multi-kernel extreme learning machine. *Energies* **2019**, *12*, 4504. [CrossRef]
8. Gururani, A.; Mohanty, S.R.; Mohanta, J.C. Microgrid protection using Hilbert–Huang transform based-differential scheme. *IET Gener. Transm. Distrib.* **2016**, *10*, 3707–3716. [CrossRef]
9. Kar, S.; Samantaray, S.R.; Zadeh, M.D. Data-mining model based intelligent differential microgrid protection scheme. *IEEE Syst. J.* **2015**, *11*, 1161–1169. [CrossRef]
10. Mishra, D.P.; Samantaray, S.R.; Joos, G. A combined wavelet and data-mining based intelligent protection scheme for microgrid. *IEEE Trans. Smart Grid* **2015**, *7*, 2295–2304. [CrossRef]
11. Ray, P.K.; Mohanty, S.; Kishor, N. Disturbance detection in grid-connected distributed generation system using wavelet and S-transform. *Electr. Power Syst. Res.* **2011**, *81*, 805–819. [CrossRef]

12. Mishra, M.; Rout, P.K. Detection and classification of micro-grid faults based on HHT and machine learning techniques. *IET Gener. Transm. Distrib.* **2018**, *12*, 388–397. [CrossRef]
13. Manohar, M.; Koley, E.; Ghosh, S. Microgrid protection under wind speed intermittency using extreme learning machine. *Comput. Electr. Eng.* **2018**, *72*, 369–382. [CrossRef]
14. Abdelgayed, T.S.; Morsi, W.G.; Sidhu, T.S. Fault detection and classification based on co-training of semisupervised machine learning. *IEEE Trans. Ind. Electron.* **2017**, *65*, 1595–1605. [CrossRef]
15. Hong, Y.-Y.; Cabatac, M.T.A.M. Fault detection, classification, and location by static switch in microgrids using wavelet transform and taguchi-based artificial neural network. *IEEE Syst. J.* **2019**, *14*, 2725–2735. [CrossRef]
16. Al Hayali, S.; Rahebi, J.; Ucan, O.; Bayat, O. Increasing energy efficiency in wireless sensor networks using GA-ANFIS to choose a cluster head and assess routing and weighted trusts to demodulate attacker nodes. *Found. Sci.* **2020**, *25*, 1227–1246. [CrossRef]
17. Rahebi, J.; Al-Shalah, M.M.S. Design, Modeling and Implementation of Multi-Function Protective Relay with Digital Logic Algorithm. *Avrupa Bilim Teknol. Derg.* **2020**, *19*, 549–565. [CrossRef]
18. Ab-BelKhair, A.; Rahebi, J.; Nureddin, A.A.M. A Study of Deep Neural Network Controller-Based Power Quality Improvement of Hybrid PV/Wind Systems by Using Smart Inverter. *Int. J. Photoenergy* **2020**, *2020*, 8891469. [CrossRef]
19. Nureddin, A.A.M.; Rahebi, J.; Ab-BelKhair, A. Power Management Controller for Microgrid Integration of Hybrid PV/Fuel Cell System Based on Artificial Deep Neural Network. *Int. J. Photoenergy* **2020**, *2020*, 8896412. [CrossRef]
20. Abed, A.H.; Rahebi, J.; Farzamnia, A. Improvement for power quality by using dynamic voltage restorer in electrical distribution networks. In Proceedings of the 2017 IEEE 2nd International Conference on Automatic Control and Intelligent Systems (I2CACIS), Kota Kinabalu, Malaysia, 21 October 2017; pp. 122–127.
21. Al-jumaili, M.; Rahebi, J.; Akbas, A.; Farzamnia, A. Economic dispatch optimization for thermal power plants in Iraq. In Proceedings of the 2017 IEEE 2nd International Conference on Automatic Control and Intelligent Systems (I2CACIS), Kota Kinabalu, Malaysia, 21 October 2017; pp. 140–143.
22. Abed, A.H.; Rahebi, J.; Sajir, H.; Farzamnia, A. Protection of sensitive loads from voltages fluctuations in Iraqi grids by DVR. In Proceedings of the 2017 IEEE 2nd International Conference on Automatic Control and Intelligent Systems (I2CACIS), Kota Kinabalu, Malaysia, 21 October 2017; pp. 144–149.
23. Sajir, H.; Rahebi, J.; Abed, A.; Farzamnia, A. Reduce power losses and improve voltage level by using distributed generation in radial distributed grid. In Proceedings of the 2017 IEEE 2nd International Conference on Automatic Control and Intelligent Systems (I2CACIS), Kota Kinabalu, Malaysia, 21 October 2017; pp. 128–133.
24. Chen, Z.; Li, C.; Sánchez, R.-V. Multi-layer neural network with deep belief network for gearbox fault diagnosis. *J. Vibroengineering* **2015**, *17*, 2379–2392.
25. Shao, H.; Jiang, H.; Zhang, X.; Niu, M. Rolling bearing fault diagnosis using an optimization deep belief network. *Meas. Sci. Technol.* **2015**, *26*, 115002. [CrossRef]
26. Wang, X.; Li, Y.; Rui, T.; Zhu, H.; Fei, J. Bearing fault diagnosis method based on Hilbert envelope spectrum and deep belief network. *J. Vibroengineering* **2015**, *17*, 1295–1308.
27. AlThobiani, F.; Ball, A. An approach to fault diagnosis of reciprocating compressor valves using Teager–Kaiser energy operator and deep belief networks. *Expert Syst. Appl.* **2014**, *41*, 4113–4122.
28. Onyango, L.A.; Waititu, A.; Mageto, T.; Kilai, M. A Hybrid Classification Model of Artificial Neural Network and Non Linear Kernel Support Vector Machine. *Int. J. Data Sci. Anal.* **2022**, *8*, 47–58. [CrossRef]
29. Deshpande, V.; Modi, P.; Sant, A. Analysis of Levenberg Marquardt-ANN based reference current generation for control of shunt active power filter. *Mater. Today Proc.* **2022**, *62*, 7104–7108. [CrossRef]
30. Irfan, M.M.; Malaji, S.; Patsa, C.; Rangarajan, S.; Hussain, S.M.S. Control of DSTATCOM Using ANN-BP Algorithm for the Grid Connected Wind Energy System. *Energies* **2022**, *15*, 6988. [CrossRef]
31. Singh, S. Condition Monitoring and Fault Diagnosis of Induction Motor using DWT and ANN. *Arab. J. Sci. Eng.* **2022**. [CrossRef]
32. Zhang, W.; Lin, Z.; Liu, X. Short-term offshore wind power forecasting-A hybrid model based on Discrete Wavelet Transform (DWT), Seasonal Autoregressive Integrated Moving Average (SARIMA), and deep-learning-based Long Short-Term Memory (LSTM). *Renew. Energy* **2022**, *185*, 611–628. [CrossRef]
33. Sahoo, S. Power and Energy Management in Smart Power Systems. *Artif. Intell. Smart Power Syst.* **2022**. [CrossRef]
34. Fahim, S.R.; Sarker, S.; Muyeen, S.; Sheikh, M.; Das, S.K. Microgrid fault detection and classification: Machine learning based approach, comparison, and reviews. *Energies* **2020**, *13*, 3460. [CrossRef]
35. Zhang, T.; Li, C.; Liang, J. A thyristor based series power flow control device for multi-terminal HVDC transmission. In Proceedings of the 2014 49th International Universities Power Engineering Conference (UPEC), Cluj-Napoca, Romania, 2–5 September 2014; pp. 1–5.
36. Candelaria, J.; Park, J.-D. VSC-HVDC system protection: A review of current methods. In Proceedings of the 2011 IEEE/PES Power Systems Conference and Exposition, Phoenix, AZ, USA, 20–23 March 2011; pp. 1–7.
37. Swetapadma, A.; Agarwal, S.; Chakrabarti, S.; Chakrabarti, S.; El-Shahat, A.; Abdelaziz, A.Y. Locating Faults in Thyristor-Based LCC-HVDC Transmission Lines Using Single End Measurements and Boosting Ensemble. *Electronics* **2022**, *11*, 186. [CrossRef]
38. Magnago, F.H.; Abur, A. Fault location using wavelets. *IEEE Trans. Power Deliv.* **1998**, *13*, 1475–1480. [CrossRef]

39. Borghetti, A.; Bosetti, M.; Di Silvestro, M.; Nucci, C.; Paolone, M. Continuous-wavelet transform for fault location in distribution power networks: Definition of mother wavelets inferred from fault originated transients. *IEEE Trans. Power Syst.* **2008**, *23*, 380–388. [CrossRef]
40. Mirjalili, S.; Mirjalili, S.; Lewis, A. Grey wolf optimizer. *Adv. Eng. Softw.* **2014**, *69*, 46–61. [CrossRef]
41. Tabakhi, S.; Moradi, P.; Akhlaghian, F. An unsupervised feature selection algorithm based on ant colony optimization. *Eng. Appl. Artif. Intell.* **2014**, *32*, 112–123. [CrossRef]
42. Sivagaminathan, R.K.; Ramakrishnan, S. A hybrid approach for feature subset selection using neural networks and ant colony optimization. *Expert Syst. Appl.* **2007**, *33*, 49–60. [CrossRef]
43. Yu, L.; Liu, H. Efficient feature selection via analysis of relevance and redundancy. *J. Mach. Learn. Res.* **2004**, *5*, 1205–1224.
44. Bennasar, M.; Hicks, Y.; Setchi, R. Feature selection using joint mutual information maximisation. *Expert Syst. Appl.* **2015**, *42*, 8520–8532. [CrossRef]
45. Hoque, N.; Bhattacharyya, D.; Kalita, J.K. MIFS-ND: A mutual information-based feature selection method. *Expert Syst. Appl.* **2014**, *41*, 6371–6385. [CrossRef]

Article

On-Line Monitoring of Shunt Capacitor Bank Based on Relay Protection Device

Yifeng Lin [1,*], Jingfu Gan [2] and Zengping Wang [1]

[1] School of Electrical and Electronic Engineering, North China Electric Power University, Beijing 102206, China
[2] State Grid Tangshan Power Supply Company, Tangshan 063000, China
* Correspondence: lyf3172@ncepu.edu.cn; Tel.: +86-158-1147-2573

Abstract: In modern power systems, the installation of a shunt capacitor bank is one of the cheapest and most widely used methods for improving the voltage profile. One shunt capacitor bank is composed of mass capacitor units and have ground, ungrounded, delta, wye connections that make configuration of capacitor banks is various. In the case of long-term operation, the failure of a single capacitor unit of a capacitor bank is likely to cause uneven voltage, which will lead to the breakdown and burning of the whole group, resulting in huge losses. The relay protection device can detect the simultaneous voltage and current of the capacitor. By utilizing these data from the relay, the abnormal state of the shunt capacitor banks at the initial stage of the fault can be found through monitoring the slight change in capacitance. Timely and early maintenance and repair would avoid capacitor bank faults and potentially greater economic losses. Capacitor banks have different connection modes. For ungrounded wye-connected capacitor banks with an unknown neutral point voltage, the capacitance parameters of each branch cannot be calculated. A parameter symmetry based on the calculation method for capacitor parameters is proposed. For long-term monitoring and observation of the capacitor capacitance value, the fault state and abnormal state of the capacitor are identified based on statistical methods. The simulation established by PSCAD verified that a relay protection device can realized an effective monitoring of the early abnormal state of the capacitor bank.

Keywords: shunt capacitor fault; equivalent balance equation; capacitance value calculation; capacitor monitoring

Citation: Lin, Y.; Gan, J.; Wang, Z. On-Line Monitoring of Shunt Capacitor Bank Based on Relay Protection Device. *Energies* **2023**, *16*, 1615. https://doi.org/10.3390/en16041615

Academic Editors: Guang Wang and Abu-Siada Ahmed

Received: 15 November 2022
Revised: 31 January 2023
Accepted: 3 February 2023
Published: 6 February 2023

1. Introduction

In the modern power system, capacitors are widely used in energy storage, voltage regulation, filtering and other scenarios due to their simple structure and limited manufacturing and maintenance costs [1]. As a reactive power supply, shunt capacitors can adjust the system voltage, improve the power quality and reduce the line loss [2]. As their implementation has increased, shunt capacitor banks have become one of the power devices with the highest failure rate. Faults in capacitor banks have caused group explosions and group damage many times [3,4], resulting in significant fluctuations in grid voltage, increasing active and reactive power losses, reducing the service life of capacitors and compromising the safety of the power grid.

The literature includes a great deal of in-depth research on the use of protection and monitoring technology of shunt capacitors to improve reliability and reduce losses. The most common means of protecting capacitors is to use different connection and voltage levels, with an emphasis on configuration protection [5–12] References [5,6] at the 500 kV voltage level, parallel compensation of substations and lines and series compensation capacitors. In addition, the protection of these capacitors is analyzed in detail, and the optimal protection configurations and scheme setting principles are given for each type of capacitor. Reference [7] calculates and analyzes the sensitivity and settings of relay protection under the various modes of the shunt capacitor banks in the 1000 kV ultra-high voltage (UHV) power system in China. Reference [8] introduces the setting principle for unbalanced protection of the H-bridge high-voltage capacitor banks. These methods provide excellent protection, but the specialized protection design and settings need to be based on different voltage levels and wiring forms, which are more complicated and costly; these need to be studied separately according to the actual working conditions of the system. Another research method relies on the development of smart substations to realize the monitoring function of shunt capacitors [13–16]. The capacitor fault monitoring system in [13] extracts synchronous voltage and current signals through specialized devices to monitor shunt capacitor banks in real time, which requires more space and is more costly. Reference [14] uses compensated negative sequence and neutral currents to locate internal faults and can locate faults that occur simultaneously in either of the two branches of a double-wye parallel capacitor bank. Reference [15] proposes a new scheme based on the unbalanced current at the neutral point. By calculating the unbalanced current distribution under all possible operating conditions, the fault severity and fault location can be identified. Reference [16] designs a very sensitive real-time capacitor-monitoring device based on calculating capacitance through the variation of LC oscillation frequency for early internal component failures of capacitor banks. This type of method mostly aims to design a dedicated monitoring device in an intelligent substation system that effectively reduce the high failure rate of capacitors and be highly universal.

Because the second method mentioned above is more economical and practical, this paper proposes a new monitoring method for shunt capacitors. The relay protection device installed on the bus can easily obtain the simultaneous voltage and branch current of the shunt capacitor bank (the voltage is equal to the bus voltage, and the branch current of the capacitor can be obtained from different CT taps). It will be very economical and convenient to only use this information to monitor the shunt capacitor banks. However, it has difficulty calculating the ungrounded wye configuration when there is no additional device to obtain the neutral voltage. In this case, the equations are solved by using the additional conditions of symmetric parameters of two branches and setting alarms according to guidelines and statistical methods that prevent capacitor faults and potential losses. An on-line monitoring method for shunt capacitor banks that is not affected by the connection method is constructed. The simulation verifies the effectiveness and feasibility of this method.

2. Principles of Shunt Capacitor Bank Monitoring

There are many types of shunt capacitor bank faults. When the internal capacitor fault is caused by overvoltage, harmonics, product defects, etc., the capacitance change value mostly occurs during the initial stage [17]. Factors such as overvoltage and harmonics cause abnormalities and failure of the individual capacitor unit in the internal series-parallel components. Damage to a single component compromises the operating state of other components; this situation gradually evolves and ultimately leads to the failure of the high-voltage shunt capacitors.

In the initial stage of capacitor fault, the abnormality of an individual unit of the shunt capacitor bank can be identified via variation in capacitance. The on-line monitoring method of the shunt capacitor bank can be realized based on the simultaneous voltage and current obtained by the relay protection device at the bus of the shunt capacitor [18].

The most common internal capacitor fault is the breakdown of internal capacitor units. There are three kinds of breakdown faults: electric breakdown, thermal breakdown and partial discharge breakdown. Electric breakdown is mainly due to the rapid breakdown of defective capacitors due to high voltage, high harmonics and other factors. Electric breakdown occurs over a short period of time, the relationship with environmental factors is small, maintenance cannot be performed in a timely manner and monitoring is of little significance. The other two kinds of breakdown develop gradually, which will produce a dielectric change before the breakdown fault. The fault can be found by measuring the capacitance.

According to the guide for the protection of shunt capacitor banks, there are three kinds of fuse protection: internal fuses, external fuses and fuseless. Internal fuses offer very effective protection. When a unit fails, the overcurrent causes the fuse to blow. If a single faulty unit is isolated, the shunt capacitor can continue to operate. However, if an internal fuse exhibits the failures shown in Figure 1b, the group capacitors with the faulty unit will be short-circuited, and the remaining capacitor groups will operate in overvoltage mode. Over a long period of time, a more serious fault would occur. Timely detection of changes can effectively avoid potential losses. If the faulty unit of an internal fuse is isolated as shown in Figure 1c, the shunt capacitor bank with a single faulty unit can still operate normally; the capacitance value changes, however, resulting in uneven voltage across the group. Long-term operation may also cause other unit faults, even serious failures. Timely detection and replacement of faulty parts can also effectively prevent the expansion of a fault.

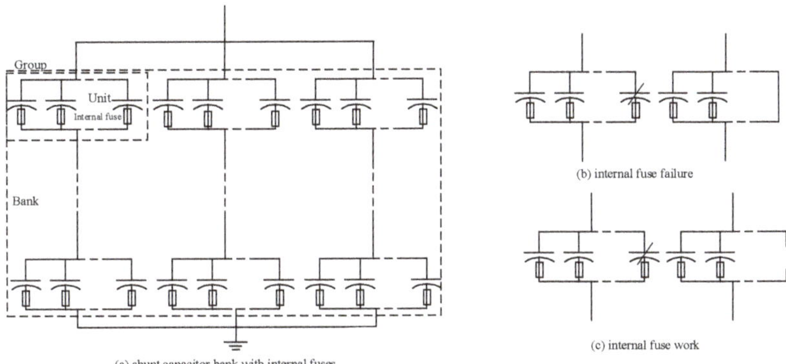

Figure 1. Shunt capacitor bank with internal fuses.

To sum up, most internal capacitor faults undergo a long process of capacitance value change. This process is affected by environment, voltage level and other factors, and it is impossible to formulate unified rules for monitoring. However, the capacitor has been comprehensively inspected at the initial stage of installation and the failure rate is low; this will be taken as the normal state at this time. By storing the data at this time and comparing real-time data with statistical methods, it can be determined whether there is a significant change. If there is a significant change, the capacitor is considered abnormal. The statistical method is general, and the reference sample is its own normal state sample; it can therefore be applied to most of the shunt capacitor banks that operate for a long time (data support).

The connections in shunt capacitor banks are wye, delta and double-wye connections. The relay protection has the following types: Zero-sequence voltage protection performs well for shunt capacitor grounding faults. Differential protection applies to all capacitor external faults. Overcurrent protection is the basic protection for all types of capacitors [19]. Double-wye connection has extra overcurrent protection at the neutral point. This connection is also divided into two cases: ungrounded and grounded [20]. The voltage at both ends of the grounding capacitor is equal to the bus voltage. The capacitance value of each phase can then be calculated by obtaining the branch current [21]. However, an ungrounded capacitor cannot calculate the capacitance value in this way, as it requires additional equipment to extract the neutral point voltage. In this paper, no additional equipment is required for calculating the ungrounded shunt capacitor bank. The simultaneous voltage and current data from the bus are the only data used to calculate the capacitance value as the monitoring criterion for the capacitor bank.

2.1. The Method of Calculating Capacitance Value

The four most common configurations of a wye-connected capacitor bank are shown in Figure 2. The CT/PT of the relay protection is built on the bus. The grounded-connected line parameters can be can easily calculated based on three-phase voltage and current, after which the capacitance change can be observed via the change in the line parameters. The calculation of ungrounded-connected line parameters is needed to obtain the voltage of the neutral point. However, increase of PT circuits means more cost and lower reliability.

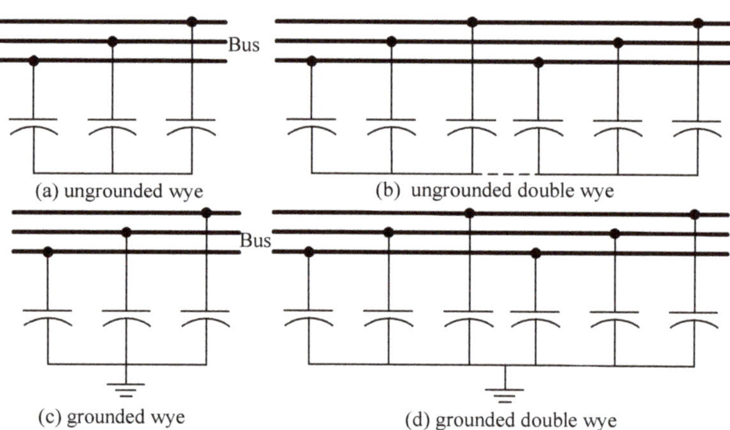

Figure 2. Most common capacitor bank configurations.

Ungrounded capacitor banks mainly consist of the wye and double-wye connections. Both wye and double-wye connections can be simplified as shown in Figure 1 [4]. We look at each phase capacitor as an impedance and we monitor its change. The change in the phase capacitance reflects the operating state of the capacitor. The advantage of monitoring the capacitance of the capacitor is that, compared with the unbalanced protection of the capacitor, the monitoring amount is the capacitance value, which is more intuitive and can better reflect the status of the capacitor. Compared with regular maintenance, monitoring is simpler and requires less time [22]. The capacitor circuit is equivalent to that shown in Figure 3, and the solution process is as follows:

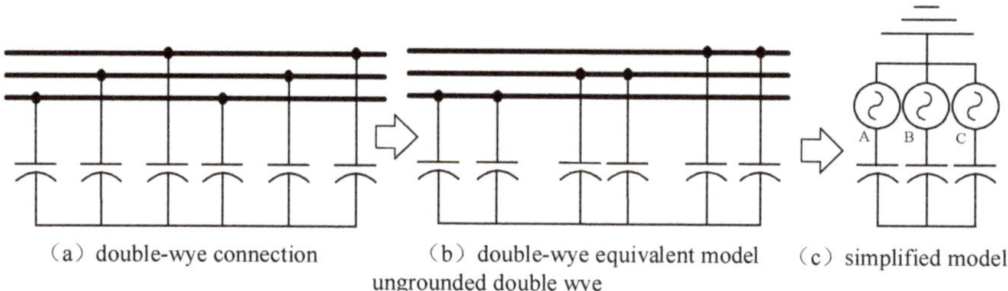

Figure 3. Capacitor equivalent circuit.

An ungrounded double-wye connection can be simplified as shown in Figure 2. The ungrounded wye and delta connections can be simplified as shown in Figure 2c. $\dot{U}_k$ $(k = A, B, C)$ and $\dot{I}_k$ $(k = A, B, C)$ are the bus voltages and branch currents of the capacitors. $\dot{Z}_A$, $\dot{Z}_B$ and $\dot{Z}_C$ are the equivalent three-phase-impedance values, including parallel capacitors, line impedances and series reactors. The two-branch-current equation is shown in the following [23].

$$\begin{cases} \dot{U}_A - \dot{I}_A\dot{Z}_A = \dot{U}_B - \dot{I}_B\dot{Z}_B \\ \dot{U}_B - \dot{I}_B\dot{Z}_B = \dot{U}_C - \dot{I}_C\dot{Z}_B \end{cases} \tag{1}$$

$\dot{Z}_A$, $\dot{Z}_B$ and $\dot{Z}_C$ are unknown quantities. Because there are only two equations in Equation (1), it cannot be solved. An additional equation is therefore required to solve the equations.

In the normal state, the three-phase parameters are equal. In the case of a single-phase fault or abnormal state, the equivalent impedance of the two other normal phases are the same. In the case of a two-unit fault in a different phase, the same working condition, model and operating conditions are assumed, so that the changes of the two-phase faults are similar. In the case of a three-phase capacitor fault, the parameters remain approximately equal for the same reason. To sum up, the operating characteristic of capacitors is that at least two-phase have the same impedances. The shunt capacitor bank must have the same three-phase capacitance when installed. Depending on the operating characteristics of the capacitor, two-phase parameters are set equal in turn, i.e., $\dot{Z}_A = \dot{Z}_B$, $\dot{Z}_A = \dot{Z}_C$ and $\dot{Z}_B = \dot{Z}_C$; these are then combined with Equation (1) to obtain Equation (2) as follows:

$$\begin{cases} \dot{U}_A - \dot{I}_A\dot{Z}_A = \dot{U}_B - \dot{I}_B\dot{Z}_B \\ \dot{U}_B - \dot{I}_B\dot{Z}_B = \dot{U}_C - \dot{I}_C\dot{Z}_B \\ \dot{Z}_A = \dot{Z}_B \text{ or } \dot{Z}_B = \dot{Z}_C \text{ or } \dot{Z}_B = \dot{Z}_C \end{cases} \tag{2}$$

When $\dot{Z}_A = \dot{Z}_B$, we obtain the solution $\begin{cases} \dot{Z}_{A1} \\ \dot{Z}_{B1} \\ \dot{Z}_{C1} \end{cases}$.

When $\dot{Z}_B = \dot{Z}_C$, we obtain the solution $\begin{cases} \dot{Z}_{A2} \\ \dot{Z}_{B2} \\ \dot{Z}_{C2} \end{cases}$.

When $\dot{Z}_B = \dot{Z}_C$, we obtain the solution $\begin{cases} \dot{Z}_{A3} \\ \dot{Z}_{B3} \\ \dot{Z}_{C3} \end{cases}$.

In order to simplify the calculation, three-phase voltages are set symmetrically and the three-phase voltages are $\dot{U}_A$, $\dot{U}_B$ and $\dot{U}_C$. Under normal working conditions, three-phase line parameters are also symmetrical. $\dot{U}_A$, $\dot{U}_B$, $\dot{U}_C$, $\dot{I}_A$, $\dot{I}_B$ and $\dot{I}_C$ therefore have the following relationship:

$$\dot{U}_B = \dot{U}_A\angle-120°,\ \dot{U}_C = \dot{U}_A\angle120° \tag{3}$$

$$\dot{I}_B = \dot{I}_A\angle-120°,\ \dot{I}_C = \dot{I}_A\angle120° \tag{4}$$

when the assumption condition is consistent with the actual situation, the solution resulting from the state equation is correct, and when the assumption condition is not consistent with the actual situation, the solution resulting from the state equation is incorrect.

Assuming that phase A fails, then analysis proceeds via the variable method. When the capacitance value of phase A decreases, the three-phase currents become $\dot{I}_A + \dot{\lambda}_a$, $\dot{I}_B$ and $\dot{I}_C$, where $\dot{I}_A$, $\dot{I}_B$ and $\dot{I}_C$ are still three-phase symmetrical, and $\dot{\lambda}_a$ and $\dot{I}_A$ are in the same direction.

① When it is assumed that phase B and phase C impedances are same (that is, the actual condition), the results are

$$\begin{cases} \dot{Z}_{A1} = \frac{1}{\dot{I}_A + \dot{\lambda}_a} (\dot{U}_A - \dot{U}_B + \dot{I}_B \frac{\dot{U}_B - \dot{U}_C}{\dot{I}_B - \dot{I}_C}) \\ \dot{Z}_{B1} = \dot{Z}_{C1} = \frac{\dot{U}_B - \dot{U}_C}{\dot{I}_B - \dot{I}_C} \end{cases} \tag{5}$$

It is known that, in this case, the calculation result is correct, the capacitance value change of phase B and phase C is 0, the impedance of phase A declines most—that is, the change amount is the largest—and $\dot{Z}_{A1}$ is used for comparison.

② When it is assumed that the parameters of phases A and B are equal (not the actual condition), the results are

$$\begin{cases} \dot{Z}_{A2} = \dot{Z}_{B2} = \frac{\dot{U}_A - \dot{U}_B}{\dot{I}_A + \dot{\lambda}_a - \dot{I}_B} \\ \dot{Z}_{C2} = \frac{1}{\dot{I}_C} (\dot{U}_C - \dot{U}_B + \dot{I}_B \frac{\dot{U}_A - \dot{U}_B}{\dot{I}_A + \dot{\lambda}_a - \dot{I}_B}) \end{cases} \tag{6}$$

Equations (3) and (4) can be substituted into (5) and (6), and the three-phase impedance values are compared in ① and ②, respectively. In this case, $\dot{Z}_{A1}$ is similar to $\dot{Z}_{C2}$, and all vectors in $\dot{Z}_{C2}$ are rotated 120 degrees clockwise; the molecular part is enlarged several times, yielding the following: $\frac{\left| \frac{1}{\dot{I}_A + \dot{\lambda}_a} (\dot{U}_A - \dot{U}_B + \dot{I}_B \frac{\dot{U}_B - \dot{U}_C}{\dot{I}_B - \dot{I}_C}) \right|}{\left| \frac{1}{\dot{I}_A} (\dot{U}_A - \dot{U}_C + \dot{I}_C \frac{\dot{U}_B - \dot{U}_C}{\dot{I}_B + \dot{\lambda}_a - \dot{I}_C}) \right|} < \frac{\left| \dot{I}_A + \frac{\sqrt{3}}{3} \dot{\lambda}_a \right|}{\left| \dot{I}_A + \dot{\lambda}_a \right|} < 1$. That means $\dot{Z}_{A1} < \dot{Z}_{C2}$. The group with the largest impedance variation is thus the actual condition solution. The ratio amplitude of $\dot{Z}_{A1}$ and $\dot{Z}_{A2}$ is $\frac{\left| \dot{I}_A + \frac{\dot{\lambda}_a \angle -30^\circ}{\sqrt{3}} \right|}{\left| \dot{I}_A + \dot{\lambda}_a \right|} < \frac{I_A + \frac{\sqrt{3}}{3} \lambda_a}{I_A + \lambda_a} < 1$, which means that $\dot{Z}_{A1} < \dot{Z}_{A2}$; in other words, the change of phase A in ① is greater than that of phase A in ②. Similarly, the change of phase A in ① is greater than that of phase B in ②.

It can be concluded from the above that the group of results with the largest variation is the correct solution when single-phase parameters change. In the case of two-phase change, the two parameters are the same, and the normal phase also can be seen as a change phase. It follows that the group of results with the largest variation is also the correct solution. In the case of the normal state or a three-phase fault, the three groups of results are the same, and all are correct solutions. As a result, the correct assumption conditions can be determined by finding the maximum change in the capacitance value, and then the correct solution of the three-phase capacitance value can be calculated.

It could calculate $\max\limits_{\substack{i = A,B,C \\ j = 1,2,3}} (\dot{Z}_{ij} - \dot{Z}_0)$, where $\dot{Z}_0$ is the initial value of single-phase impedance, and find the max impedance change. Then, this group ($\dot{Z}_{Aj}$, $\dot{Z}_{Bj}$ $\dot{Z}_{Cj}$) is determined as correct calculation. If the parameters of the series reactor are defined as L, and the system frequency is f, then the results of the three-phase capacitance values C_A,C_B and C_C are as follows:

$$\begin{cases} C_A = \frac{1}{4\pi^2 f^2 L - 2\pi f Im(\dot{Z}_{Aj})} \\ C_B = \frac{1}{4\pi^2 f^2 L - 2\pi f Im(\dot{Z}_{Bj})} \\ C_C = \frac{1}{4\pi^2 f^2 L - 2\pi f Im(\dot{Z}_{Cj})} \end{cases} \tag{7}$$

2.2. Monitoring Criteria

After solving for the capacitance value, taking into account the normal fluctuations and calculation errors, and in accordance with IEEE Guide for the Protection of Shunt Capacitor Banks [24], the shunt capacitor is considered faulty when the calculated capacitance value

C_k and the rated value C_N do not meet $|C_k - C_N| \times 100\% > 5\%$. Repair is required. Cst is the statistical data based on the stored capacitance from the protection device. When there is a significant difference between the current capacitance value and the past sample ($C_A \notin C_{st}$), an alarm is issued. Because relay protection is sensitive to capacitor external fault, and a bus fault would make the voltage zero, the shunt capacitor bank monitor should not trigger an alarm due to an external fault or bus fault. Hence, in order to avoid a monitoring malfunction, a low voltage criterion is added. When a capacitor external fault or bus fault occurs, the voltage will drop significantly. As a result, when any voltage of the three-phase bus is $U_k < 0.85U_N$, this phase is determined to be a short-circuit fault, the protection will trip and capacitor monitoring does not need an alarm. Considering the need to prevent disturbance, a certain delay is added. The monitoring logic is shown in Figure 4.

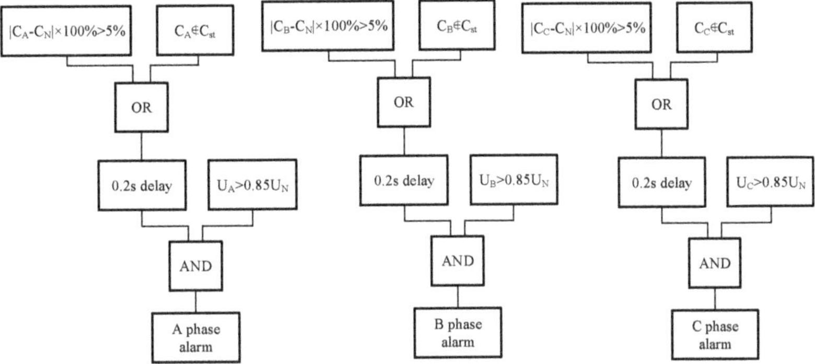

Figure 4. Monitoring logic.

Considering that capacitor values are affected by environmental factors such as operating temperature, air pressure, dust, etc., adding a year-on-year comparison of capacitance values can reduce the influence of operating conditions on measured values and indicate whether the capacitor has changed significantly after long-term operation; doing so can improve the sensitivity and accuracy of monitoring.

The normal state data for the current and previous years are stored, two sets of capacitance values from the same month are sampled and a paired-sample t test is performed to check for significant differences; if there is a significant difference, the operating state is considered abnormal and an alarm is issued.

First, this assumes that the mean capacitance values of the two months are the same, i.e., there is no significant difference.

Second, the formula for calculating t is given below. X_1/X_2 are the sample data from the previous and current year for the same month. $\overline{X_1}/\overline{X_2}$ are the averages of the two samples. $\delta_{X_1}/\delta_{X_2}$ are the variances of the two samples. γ is the correlation coefficient of the two samples.

$$t = \frac{\overline{X_1} - \overline{X_2}}{\sqrt{\frac{\delta_{X_1}^2 + \delta_{X_2}^2 - 2\gamma\delta_{X_1}\delta_{X_2}}{n-1}}} \tag{8}$$

Third, we assume a confidence level of 95% and $t(29)_{0.05} = 2.045$ according to the T value table. If $t \leq 2.045$, no significant difference is found, and the capacitor is operating normally. If $t > 2.045$, a significant difference is found, and the capacitor is operating abnormally.

3. Analysis of Test Results

To verify the practicability of the monitoring method in this paper, the system is simulated and analyzed by PSCAD/EMTDC simulation software. Figure 5 shows the simulated circuit diagram.

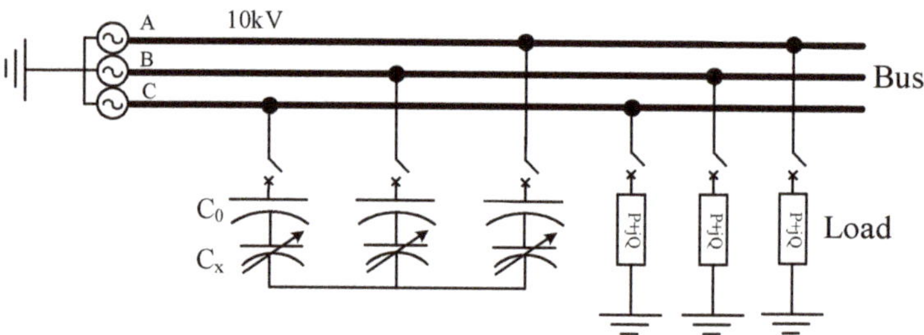

Figure 5. Shunt capacitor bank at 10 kV bus.

The simulation system is a shunt capacitor bank built on a 10 kV bus. C0 is the initial capacitance and CX is the variable capacitance, i.e., the simulated capacitance change in the abnormal state. The capacitor branch resistance is 10 Ω, the capacitance value is 9.76 μF and the series reactor is 63.94 mH, given 5% of capacitance.

3.1. Capacitor Internal Fault

Figure 6 shows the calculation results when phase A capacitance declined by 10% of the standard value. The bus voltage remained unchanged. Phase A RMS current varied from 20.26 A to 18.96 A. Phase B/C RMS current varied from 20.26 A to 19.94 A. As can be seen from the figure, the capacitance changes slightly, the power system still operates normally and the voltage and current do not change significantly. It is correctly identified that the variability of phase A capacitance is out of specification and a phase A alarm is issued. The calculation result of the capacitance is 8.7919 μF; the actual capacitance is 8.784 μF (with less than 0.1% relative error).

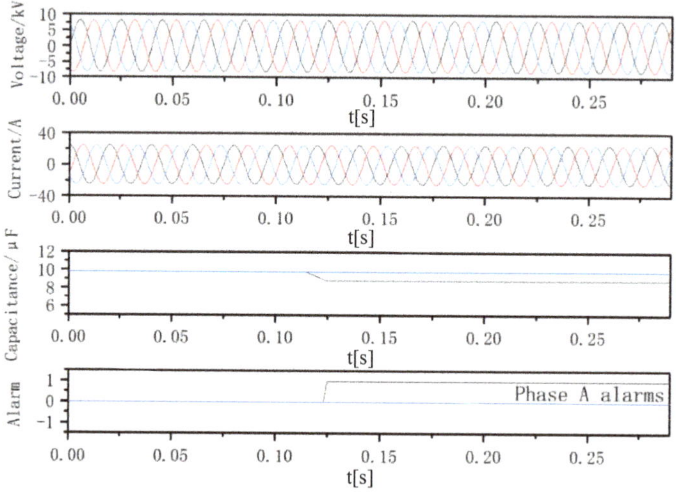

Figure 6. Phase A internal fault.

Figure 7 shows the simulation when the phase A and B capacitances declined by 10% of the standard capacitance value. The bus voltage remained unchanged. Phase A/B RMS current varied from 20.26 A to 18.55 A. Phase B/C RMS current varied from 20.26 A to 19.54 A. As can be seen from the figure, this method can correctly detect that the capacitance changes of phase A and phase B are out of specification and issue an alarm. The calculated capacitance value for phase A/B is 8.7972 μF; the actual capacitance is 8.784 μF (with only 0.15% relative error).

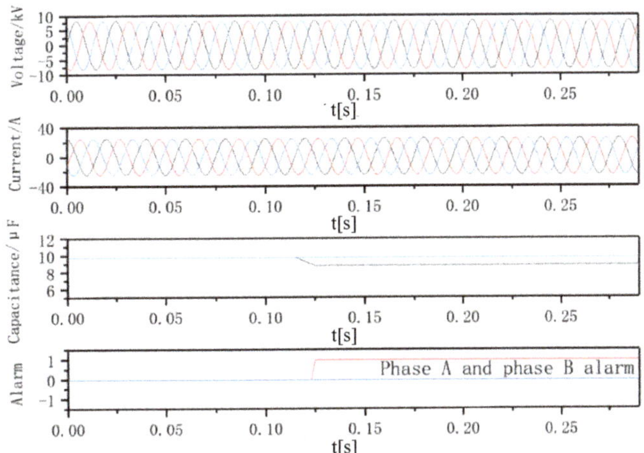

Figure 7. Phase A and B internal fault.

Figure 8 shows the simulation when the capacitances of three phases change at the same time. (A, B and C declined 2%, 4% and 8%, respectively, compared with the standard value). The bus voltage remained unchanged. The phase A RMS current varied from 20.26 A to 19.57 A. The phase B RMS current varied from 20.26 A to 19.37 A. The phase C RMS current varied 20.26 A to 18.97 A. It can be seen from the figure that when there is a deviation in the initial parameters, capacitor monitoring can correctly detect that the phase C capacitance change exceeds specified values and a phase C alarm is triggered. The calculated result of the capacitances of phase A and B are 9.4608 μF (the actual capacitance values are 9.5648 μF and 9.3696 μF, the relative error is 1.09% and 0.97%). The calculated result of phase C capacitance is 8.9873 μF, the actual capacitance is 8.9792 μF and the relative error is 0.09%. There is a certain error in the calculated results of the normal phases A and B due to the initial deviation, but within the acceptable range, the capacitance value of the faulty phase C is still accurately calculated, the abnormality is correctly identified, and an alarm is issued.

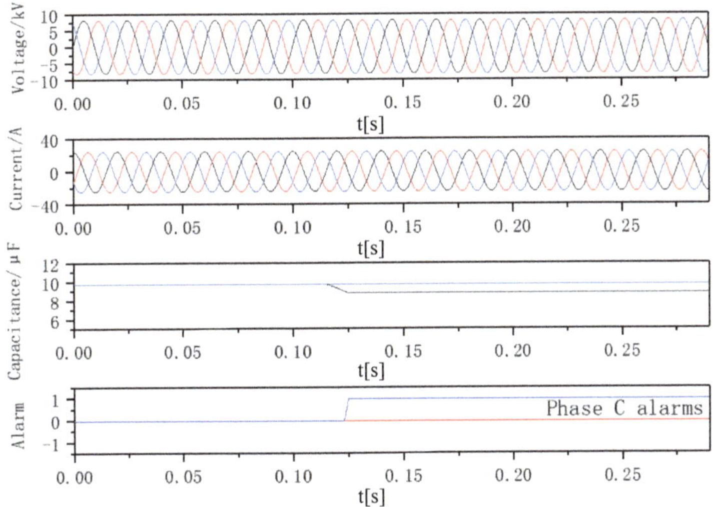

Figure 8. Phase C internal fault.

The above results show that when some capacitor units are faulty, the slight change of capacitance value can be accurately detected. When the capacitance value changes beyond the shunt capacitor guidelines, or a statistical data comparison flags the value as abnormal, an alarm will be issued, which is convenient for maintenance.

3.2. External Fault

External faults occur when the fault accrues on a bus or line. The relay protections operate in these cases. The bus voltage almost drops to zero, and the capacitance value calculation is meaningless in this situation. In addition, capacitor monitoring should not trigger an alarm at external fault. Figure 9 shows the normal operation of the bus in the case of a single-phase ground fault. It can be seen from the figure that the capacitance value before and after the fault has a bump, because when calculating the current and voltage, a short circuit in one cycle causes a sudden change during the fault. The capacitance value is stable, and its calculated value is 208.23 μF, because when an external fault occurs, the measured impedance is the impedance from the measuring point to the short circuit point. The external fault should be tripped by the relay protection, and the capacitor monitoring should not trigger an alarm. Adding a low-voltage block make the alarm would not malfunction during the external fault.

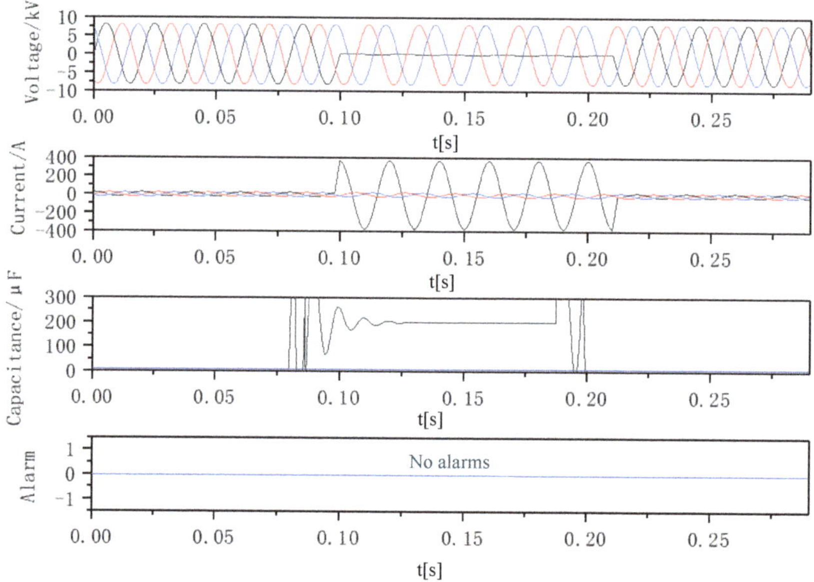

Figure 9. Phase A ground fault simulation results.

Figure 10 shows the normal operation of the bus in the case of phase A and B grounding faults. During the fault, the calculated phase A and phase B capacitance values are both 208.23 μF; A, B two-phase low-voltage criteria are activated; two-phase capacitor failure monitoring is no longer active; and no malfunction occurred.

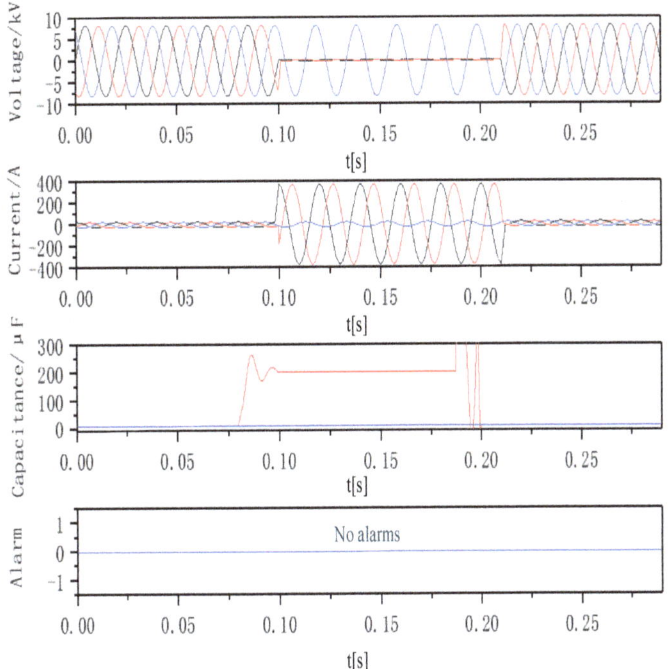

Figure 10. A and B grounded external fault.

The above results show that the capacitor fault alarm does not malfunction, and the fault is removed by the protection mechanism.

Summarizing various faults of capacitors as shown in Appendix A, it can be seen that this method can correctly detect the internal faults and locate the abnormal phase. This is convenient for maintenance and repair work.

4. Conclusions

The on-line monitoring method for shunt capacitors proposed in this paper has the following characteristics:

(1) This monitoring method is applicable to shunt capacitor banks of all connection types. It can be realized only by using a relay protection device, with no additional device to measure the state quantity. The method is economical and convenient.

(2) In the event of a slight capacitor failure or abnormality, the abnormal phase can be detected to maintain safe operation.

Author Contributions: Conceptualization, Y.L. and Z.W.; methodology, Y.L.; software, Y.L.; validation, J.G.; formal analysis, Z.W.; investigation, J.G.; resources, J.G.; data curation, J.G.; writing—original draft preparation, Y.L.; writing—review and editing, Y.L.; visualization, J.G.; supervision, Z.W.; project administration, J.G.; funding acquisition, J.G. All authors have read and agreed to the published version of the manuscript.

Funding: This research received no external funding.

Data Availability Statement: The data presented in this study are openly available in IEEE, at references [13,24].

Conflicts of Interest: The authors declare no conflict of interest.

Appendix A

Table A1. Alarm results in each capacitor fault.

Fault Conditions	Fault Phase Impedance			Alarm			Result Analysis
	A	B	C	A	B	C	
A-phase capacitor failure	abnormal	normal	normal	Yes	No	No	Capacitor failure, faulty phase correct alarm
Bphase C capacitor failure	normal	abnormal	abnormal	No	Yes	Yes	Capacitor failure, faulty phase correct alarm
Three-phase capacitor failure	abnormal	abnormal	abnormal	Yes	Yes	Yes	Capacitor failure, faulty phase correct alarm
A-phase short circuit	abnormal	normal	normal	No	No	No	External fault does not alarm, the signal is correct
Bphase C short circuit	normal	abnormal	abnormal	No	No	No	External fault does not alarm, the signal is correct
Three-phase short circuit	abnormal	abnormal	abnormal	No	No	No	External fault does not alarm, the signal is correct

References

1. Kinjo, T.; Senjyu, T.; Urasaki, N.; Fujita, H. Output levelling of renewable energy by electric double-layer capacitor applied for energy storage system. *IEEE Trans. Energy Convers.* **2006**, *21*, 221–227. [CrossRef]
2. Aziz, M.M.A.; Abou El-Zahab, E.E.D.; Ibrahim, A.M.; Zobaa, A.F. Effect of connecting shunt capacitor on nonlinear load terminals. *IEEE Trans. Power Deliv.* **2003**, *18*, 1450–1454. [CrossRef]
3. Panda, N.R.; Pachpund, S. Capacitor Bank Balancing: Causes and Practical Levels of Unbalance. *IEEE Ind. Appl. Mag.* **2022**, *28*, 12–19. [CrossRef]
4. Pour, M.R.; Azimian, M. Analysis of high voltage shunt capacitor bank over-voltage breakdown detection. *Future Energy* **2022**, *1*, 16–23. [CrossRef]
5. Wei, J.; Lan, J.; Jiang, P. MRFO Based Optimal Filter Capacitors Configuration in Substations with Renewable Energy Integration. In Proceedings of the 2022 4th Asia Energy and Electrical Engineering Symposium, Chengdu, China, 25–28 March 2022; pp. 328–333.
6. Sun, H.; Li, H.; Yang, S. Research on the Analysis of Unbalanced Current Alarm Factors and Solution Measures Based on Fixed Series Compensation Devices. In Proceedings of the 2022 4th International Conference on Power and Energy Technology (ICPET), Xining, China, 28–31 July 2022; pp. 369–374.
7. Tian, Q.; Zhu, T. Shunt capacitor bank protection in UHV pilot project. In Proceedings of the 2016 China International Conference on Electricity Distribution (CICED), Xi'an, China, 10–13 August 2016; pp. 1–5.
8. Mei, N.; Li, Y.; Duan, X. Study on high voltage capacitor unbalance protection in HVDC projects. In Proceedings of the 2009 Asia-Pacific Power and Energy Engineering Conference, Wuhan, China, 28–30 March 2009; pp. 1–4.
9. Esponda, H.; Guillen, D.; Vazquez, E. Energy modes-based differential protection for Shunt capacitor banks. In Proceedings of the 15th International Conference on Developments in Power System Protection, Liverpool, UK, 9–12 March 2020; pp. 1–6.
10. Jouybari-Moghaddam, H.; Sidhu, T.; Parikh, P. Enhanced fault location method for shunt capacitor banks. In Proceedings of the 2017 70th Annual Conference for Protective Relay Engineers, College Station, TX, USA, 3–6 April 2017; pp. 1–11.
11. Lertwanitrot, P.; Ngaopitakkul, A. Application of Magnitude and Phase Angle to Boundary Area-Based Algorithm for Unbalance Relay Protection Scheme in 115-kV Capacitor Bank. *IEEE Access* **2021**, *9*, 35709–35717. [CrossRef]
12. Jouybari-Moghaddam, H.; Sidhu, T.S. A study of capacitor element failures in high voltage Shunt Capacitor Banks. In Proceedings of the 2017 IEEE 30th Canadian Conference on Electrical and Computer Engineering, Windsor, ON, Canada, 30 April–3 May 2017; pp. 1–4.
13. Zhang, M.; Du, J.; Gao, B. Research on the implementation scheme of shunt capacitor protection and monitoring. In Proceedings of the 2017 China International Electrical and Energy Conference, Beijing, China, 25–27 October 2017; pp. 356–359.
14. Pavan, P.S.; Das, S. Novel method for location of internal faults in ungrounded double wye shunt capacitor banks. *IEEE Trans. Power Deliv.* **2020**, *36*, 899–908. [CrossRef]
15. Shilong, L.; Yufei, T.; Mingzhong, L. An Novel On-line Monitoring Method for Double-Y Type Shunt Capacitor Bank. In Proceedings of the 2020 Asia Energy and Electrical Engineering Symposium, Chengdu, China, 28–31 May 2020; pp. 296–300.
16. Xiaoyu, C.; Jianyong, Z.; Jun, M. Power capacitor banks failure warning method based online intelligence LC oscillation frequency variation and its implementation. *Power Syst. Prot. Control.* **2015**, *43*, 144–149.
17. Jianjun, Z.; Honghua, S. Research of On-line Monitoring System for High Voltage Shunt Capacitors. *Power Capacit. React. Power Compens.* **2013**, *34*, 22–27.
18. Goodarzi, A.; Allahbakhshi, M. Online condition monitoring algorithm for element failure detection and fault location in double wye shunt capacitor banks. *Int. J. Electr. Power Energy Syst.* **2022**, *137*, 107864. [CrossRef]

19. Chi, D.; Tang, H.; Chang, H. Failure analysis of fuse for external protection of capacitor bank. In Proceedings of the 18th International Conference on AC and DC Power Transmission (ACDC 2022), Windsor, ON, Canada, 2–3 July 2022; pp. 185–188.
20. Jena, S.; Mohanty, R.; Pradhan, A.K. A traveling wave based method for protection of shunt capacitor bank. IEEE Transactions on Power Delivery. *IEEE Trans. Power Deliv.* **2021**, *37*, 2599–2609. [CrossRef]
21. Mohanty, R.; Pradhan, A.K. Fast and Sensitive Time-Domain Protection of Shunt Capacitor Banks. In Proceedings of the 2021 9th IEEE International Conference on Power Systems (ICPS), Kharagpur, India, 16–18 December 2021; pp. 1–6.
22. Bastos, A.F.; Santoso, S. Condition monitoring of circuit switchers for shunt capacitor banks through power quality data. *IEEE Trans. Power Deliv.* **2019**, *34*, 1499–1507. [CrossRef]
23. Horton, R.; Warren, T.; Fender, K. Unbalance protection of fuseless, split-wye, grounded, shunt capacitor banks. *IEEE Trans. Power Deliv.* **2002**, *17*, 698–701. [CrossRef]
24. IEEE Guide for the Protection of Shunt Capacitor Banks—Redline (*C37.99-2012—Redline*). *IEEE Std.* **2013**, 1–299.

Article

Condition-Based Maintenance of Gensets in District Heating Using Unsupervised Normal Behavior Models Applied on SCADA Data

Valerio Francesco Barnabei [1,*], Fabrizio Bonacina [1], Alessandro Corsini [1], Francesco Aldo Tucci [1] and Roberto Santilli [2]

[1] Department of Mechanical and Aerospace Engineering, University of Rome La Sapienza, Via Eudossiana 18, I00184 Rome, Italy
[2] ENGIE Servizi S.p.A, District Heating and Power, Viale Avignone 12, I00144 Rome, Italy
* Correspondence: valerio.barnabei@uniroma1.it

Abstract: Increasing interest in natural gas-fired gensets is motivated by District Heating (DH) network applications, especially in urban areas. Even if they represent customary solutions, when used in DH, duty regimes are driven by network thermal energy demands resulting in discontinuous operation, which affects their remaining useful life. As such, the attention on effective condition-based maintenance has gained momentum. In this paper, a novel unsupervised anomaly detection framework is proposed for gensets in DH networks based on Supervisory Control And Data Acquisition (SCADA) data. The framework relies on multivariate Machine-Learning (ML) regression models trained with a Leave-One-Out Cross-Validation method. Model residuals generated during the testing phase are then post-processed with a sliding threshold approach based on a rolling average. This methodology is tested against nine major failures that occurred on the gas genset installed in the Aosta DH plant in Italy. The results show that the proposed framework successfully detects anomalies and anticipates SCADA alarms related to unscheduled downtime.

Keywords: multivariate time series; early fault detection; condition based maintenance; multi-MW gensets SCADA data

Citation: Barnabei, V.F.; Bonacina, F.; Corsini, A.; Tucci, F.A.; Santilli, R. Condition-Based Maintenance of Gensets in District Heating Using Unsupervised Normal Behavior Models Applied on SCADA Data. *Energies* **2023**, *16*, 3719. https://doi.org/10.3390/en16093719

Academic Editors: Shunli Wang, Jiale Xie and Guang Wang

Received: 14 March 2023
Revised: 14 April 2023
Accepted: 19 April 2023
Published: 26 April 2023

1. Introduction

District Heating, also known as heat networks or teleheating, provides a platform for heat supply based on the integration of low-carbon technologies, including renewable energy sources and thermal storage, to improve overall efficiency and minimize greenhouse gas emissions. In operation since the end of the XIX century, DH represents an efficient way to provide heat to a large number of users in densely populated urban areas [1–5]. According to IEA's 2021 report [6], DH systems are important solutions to describe the heating sector in any NZE 2050 scenario [7].

DH systems are composed of thermal plants and a distribution network of insulated pipes that deliver heat to the end users. The thermal plant is based on technology to generate heat from fossil fuels or renewable energy sources or to valorize waste heat [8]. In 2020, nearly 90% of heat was produced from fossil fuels, and one of the most common technologies in DH thermal power plants involves the use of generator sets, also known as gensets, with internal combustion engines (ICEs) either in combined heat and power (CHP) configurations or directly coupled with heat pumps [9].

Wang et al. [10] reported that, in 2012, in China, more than 36% of the total building energy demand was consumed for residential heating purposes, and about 62.9% of district heat was produced by CHP systems. As another example, in Finland, DH accounts for about 50% of the total heating market, and the city of Helsinki has around 20% of their district heat produced by genset with the use of wastewater as a low-grade heat source [11].

Gensets can suffer from intermittent operation caused by the variability and seasonality of the network heat demand, especially when directly coupled with heat pumps. These operation modes often lead the engine off-design and can be interpreted as the root cause of genset anomalies and failures. Therefore, the research on automatic Fault Detection (FD) of gensets based on proper Condition-Based Maintenance (CBM) strategies is of paramount importance to monitor the operation, reduce downtime and ensure the reliability and productivity of the overall heat supply process [12–14].

Rooted in condition-monitoring systems, CBM aims to establish frameworks for the diagnosis of equipment under supervision indicating incipient failures using sensor networks. CBM defines and monitors health indicators capable of signaling an anomaly in the case of deviation from reference values. Based on the evaluation of the current state of the equipment, it is possible to identify faults and malfunctions at an early stage, thus, allowing the timely planning of maintenance interventions. Despite the fact that scheduled maintenance and CBM are complementary, CBM is, by far, the most cost-effective approach and the one that enhances the life expectancy of the equipment [15,16].

A recent review on ICE diagnostics [17] suggested that a limited number of papers dealt with analytical models specifically designed for the CBM of gensets operating in DH networks. Most of the literature is dedicated to load prediction and the analysis of optimal network design with few contributions focusing on the operation and maintenance of networks and distribution pipelines [18].

As reported in [19], Machine-Learning (ML) algorithms have also been established as a viable solution in the DH scenario because they are easily adaptable to changing conditions, capable of modeling non-linear phenomena and can benefit from the historical data readily available in modern control systems (e.g., SCADA data). While ML approaches based on classification algorithms, such as Bayesian Classifiers (BCs) or Support Vector Machines (SVMs), have been widely used for FD of ICEs [20–25], regression algorithms seem to represent the most suitable option to perform an effective CBM.

In fact, on the one hand, BCs and SVMs are supervised ML tools that enable effective FD, but they rely on events that already occurred in the past to label the training dataset. On the other hand, unsupervised models based on regression approaches, classified in [26] as Normal Behavior Models (NBM), are able to detect anomalies in real-time conditions, as they can signal upcoming fault events in advance.

As a general outline, NBM approaches for CBM consist of training a reference model that represents the normal operation of the system and evaluating the deviation, or residual, between the predicted and actually measured values in real-time conditions to detect anomaly occurrence. Note that training a regression model to create an NBM may appear to be a supervised approach because it is trained on examples in which the expected values of the target variable are also provided; however, due to the absence of labels classifying the operational state in the training phase, NBM models fall into the category of unsupervised fault-detection methods [27].

The scope of this work is to propose an unsupervised NBM model designed for gensets operating in DH networks that introduces a series of advantages with respect to the state of art as detailed in the following section.

2. Unsupervised CBM of ICEs: State of the Art

To date, most applications of data-driven unsupervised fault detection in ICE fall in the automotive, aviation and marine sectors. To name a few, Liu et al. [28] used a linear regression based on thermal and electrical parameters for detecting the valve clearance of diesel engines. Bryght et al. [29] predicted failure in aircraft engines by combining lead function and logistic regression applied to aircraft engine takeoff data. Singh et al. [30] tested the performance of several Machine-Learning algorithms for predicting the health of an aircraft engine on historical data retrieved from the NASA data repository. Maraini et al. [31] developed a data-driven framework based on a Multi-Layer Perceptron (MLP) for marine gas turbine engine health monitoring. Chen et al. [32] proposed a deep autoencoder with a Dimension Fusion Function method (DFF-DAE) to detect aero-engine faults.

Focusing on the specific applications of ICEs in power plants, Mendonça et al. [12] proposed a methodology for the detection of incipient failures in the components of internal combustion engine-driven generators based on Electrical Signature Analysis (ESA), while Deon et al. [33] introduced a predictive maintenance module within a digital twin based on the definition of independent subsystems, each one supported by an ad hoc trained model (Air Intake Subsystem, Exhaust Subsystem, Fuel Subsystem, Water Cooling Subsystem, Lubrication Subsystem and Mechanical Subsystem).

Based on the above, it can be concluded that a large part of the literature envisages the development of different Machine-Learning models applied to data sampled from sensor networks specifically designed for condition-monitoring systems (e.g., accelerometers and vibration sensors). On the other hand, as an interesting perspective, in recent years, there has been an increasing focus on NBM approaches for CBM based on SCADA data, especially in the context of wind turbines (see [26] for a comprehensive review).

However, NBM approaches can present a number of critical issues when applied to multivariate SCADA data. In this sense, a number of challenges were identified in [27]. As a first example, the high data dimensionality heavily affects the response times of NBM models, making them frequently unsuitable for near real-time applications typical of CBM. A second concern is represented by the challenge in isolating the size of the time window to train the reference model: the seasonal nature of the operating conditions, coupled with the possible presence of undesired anomalies in the dataset, makes it difficult to identify the standard dynamics of the system using, for example, standard approaches for clustering or outlier isolation.

Finally, a further issue is represented by the appropriate handling of residuals for alarm activation. Since residuals are evaluated as the difference between the value of a signal predicted by the regression model (trained under reference conditions) and the actual value of the same signal logged by the SCADA sensor, they can present a high level of noise and typical signal variability, which makes it very challenging to trigger alarms using standard control charts.

As an attempt to face the aforementioned issues and challenges, a general framework for SCADA-based CBM using a NBM approach is proposed, and the method is applied to the technology of natural gas (NG) gensets in DH networks. Specifically, the framework proposes a series of solutions to manage the entire data-mining process, starting from the reduction of dimensionality in the pre-processing phase with a feature-selection algorithm, passing through the training methods of the reference models with a Leave-One-Out Cross-Validation approach [34], up to the post-processing of residuals by means of the introduction of a two-stage sliding threshold metric to provide nowcasting of the alarms. For the ML module, two different regression algorithms, namely, XGB and MLP, are trained and compared.

The framework is tested on SCADA data sampled on a 7.5 MW NG genset installed in the District Heating plant of the city of Aosta, Italy. The considered dataset includes 45 parameters with 5 min sampling during 16 months of engine operation (from September 2019 to December 2020). The paper is organized as follows. Section 3 presents the discussion of the building blocks of the proposed ML framework for CBM. Then, Section 4 describes the case study and the obtained results. Finally, Section 5 summarizes the present work and presents our conclusions.

3. Anomaly Detection Framework, Overview

The first operation proposed in the framework is the pre-processing and cleaning of SCADA event logs and signals, filtering out minor events from the logs and removing constant signals (see Section 3.1).

Subsequently, we process all SCADA signals with a feature-selection method based on a variable importance approach to select the best predictors for the nowcasting of a specific target variable (see Section 3.2). These preliminary operations optimize the performance of the ML models both in terms of accuracy and computational costs for CBM purposes.

In the next step, we apply two completely different models (namely, XGB and MLP) independently for the construction of the reference model, training both of them with a Leave-One-Out Cross-Validation approach (see Section 3.3). This avoids any risk of overfitting and guarantees greater robustness and flexibility of the results by simulating unsupervised real-time applications.We recommend having at least one year of data for the training phase, to guarantee the effective learning of the recurring relational dynamics between signals while still taking into account the seasonal operational variations typical of the analyzed users.

At the testing stage, we adopt a warning rule for anomaly detection based on a sliding threshold metric approach, applied to the Local Residual Indicators (LRIs) of each parameter. Specifically, we filter the noise of LRIs and subsequently define a control chart based on their intensity and time persistence to trigger alarms only related to significant anomalies and to reduce the occurrence of false positives (see Section 3.5).

Finally, we evaluate the anomaly detection results with respect to the ability to identify precursors from the SCADA event logs and early detect major faults. Concerning the SCADA event logs, after a preliminary filtering of minor events, the framework integrates the evaluation of the Mean Time Between Alarms (MTBA) indicator and the quantification of the total downtime in a prognostic perspective.

The entire framework is implemented using Python 3.9 Scikit-Learn open-source library [35]. A step-by-step framework description is given in the following Figure 1.

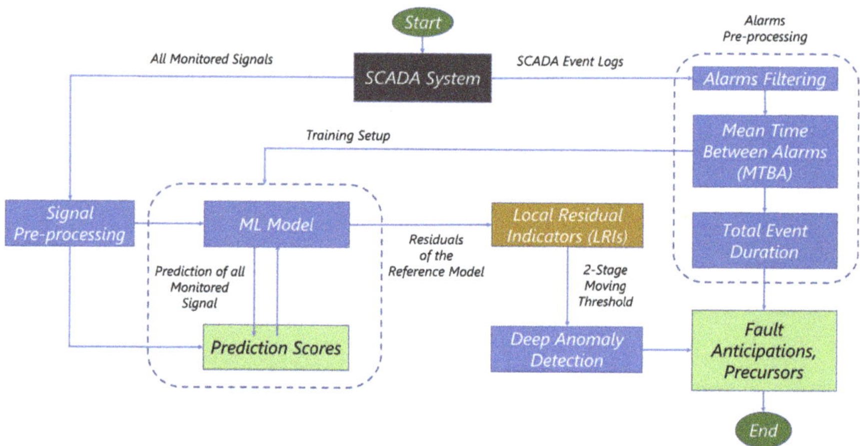

Figure 1. ML framework for CBM and schematics.

3.1. SCADA Event Log and Signal Pre-Processing

The pre-processing of the SCADA event logs filters all minor alarms unrelated to specific faults or anomalies, along with events recorded during the engine downtime. The remaining logs are then used to estimate operation metrics, such as the MTBA and the total duration of the outage events until correct operations are recorded. Those indicators represent key parameters for the training setup of the ML model (as explained in more detail in Section 3.4). Additionally, we evaluated the information content of each signal time series using the Shannon Entropy (H) metric [36], which allows for the interpretation of parameters with H close to zero as irrelevant or derived and to remove them from the training dataset, together with constant signals. Finally, a sigma rule was adopted to identify and remove extreme outliers related to measurement errors and to finally filter the signals with respect to the active power of the ICE.

3.2. Feature Selection

The framework adopts a feature-selection method based on variable importance through exploiting the Predictive Power Score (*PPS*) [37] algorithm. The output of the *PPS* analysis is an asymmetric, data-type independent index that identifies the relationships among the features in a dataset. Specifically, *PPS* quantifies how much a single input variable affects the prediction of the target variable. *PPS* assigns an index on each single input feature (x_i) at a time used to predict the target variable (y_i) via a Decision Tree algorithm. The index is expressed as:

$$PPS = 1 - \frac{MAE^{x_i,y_i}_{model}}{MAE^{y_i}_{naive}} \tag{1}$$

where MAE^{x_i,y_i}_{model} is the Mean Absolute Error of the chosen regression model that predicts y_i from a candidate x_i, while $MAE^{y_i}_{naive}$ is obtained with a naive model that always predicts the median of y_i. The index ranges from 0 (no predictive power) to 1 (perfect predictive power). On this basis, as suggested by the authors of the algorithm [37], the minimum *PPS* acceptability limit is consistently set at 0.2. For each specific target variable (y_i), a vector of best predictors B_i is defined, selecting from the set of all possible input features (x_i), the ones with a *PPS* score above the set threshold. For example, as highlighted in Figure 2, for the specific target variable (y_i) the vector of best predictors B_i includes the subset of input features ranging from x_1 to x_8.

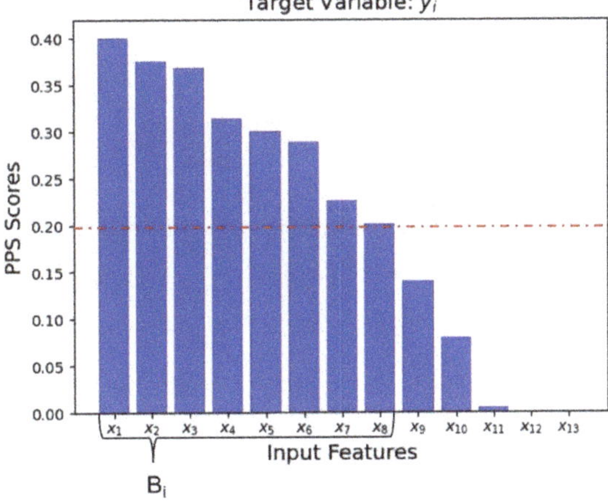

Figure 2. Example of criteria used to select best predictors based on the PPS score. Bars represent the score value, while the red dashed line represent the minimum acceptability limity for the score.

3.3. Machine-Learning Model

Two different regression algorithms, namely, XGB and MLP, are selected as candidates for the ML module. Both the regression algorithms saw an optimization process using a grid search approach [38] to select the best combination of hyper-parameters. In spite of the fact that both models identify within the training dataset one parameter at a time as the target variable (y_i) and exploit all the others to predict it, some core differences between the models still represent a challenge for comparability.

Notably, since XGB belongs to the category of ensemble algorithms and since its structure is composed of several decision trees, the results are independent from feature normalization [39]. In contrast, Artificial Neural Networks rely on statistical analysis and, thus, are strongly influenced by the distribution and quality of the data and are highly dependent on the order of magnitude of their input values. As a consequence, MLP may neglect or overestimate the influence of certain features according to their values [40]. To avoid this, input signals are initially normalized for the MLP model using a Standard Scaler and then the predicted features are scaled back to their original size.This ensures the comparability of results between the two ML models in terms of prediction scores.

3.4. Training Setup

As previously stated, the training strategy relies on a Leave-One-Out Cross-Validation method [34] as a proposed solution to isolate reference operating conditions with standard unsupervised approaches in highly discontinuous duty periods combined with the strong seasonality of the signals. In the specific DH application presented in the paper, the genset workload presented strong discontinuities in the summer period as well as a higher environmental temperature operating condition, while having a more continuous workload in winter with lower external temperatures.

In detail, as shown in Figure 3, one month m is cyclically isolated as the testing dataset D_{test}, and a model is trained on the remaining months split between training D_{train} and validation D_{val} datasets. This approach is meant to avoid possible overfitting and presume that most of the operational data over a long period of time refers to normal engine operation. To further reduce the possible presence of failure precursors in the reference model, D_{train} does not include any downtime period, considering an additional safety time range equal to the value of the MTBA index obtained at the pre-processing stage.

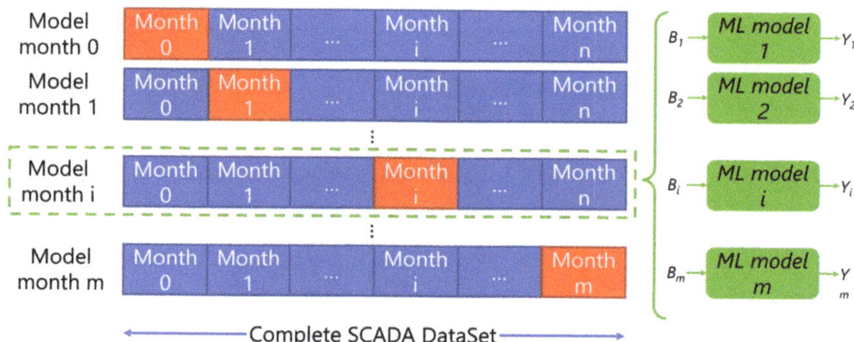

Figure 3. Representation of the Leave-One-Out Cross-Validation method as implemented in the present study.

As a result of this training process, a specific regression model (ML model$_i$) for each target variable (y_i) is obtained and defined as a function of best predictors B_i previously identified. The accuracy of the two models during the training phase on the reference period was evaluated with customary scores, i.e., the Mean Absolute Error (MAE), Mean Squared Error (MSE), Root Mean Squared Error (RMSE) and Mean Absolute Percentage Error (MDAPE).

3.5. Residual Indicator Definition

The aim of the proposed CBM framework is the definition of anomaly detection rules to trigger early warnings of incipient failures. To this end, an LRI is defined for each monitored variable [41] as the absolute value of the difference between the actual values (f) and those predicted by the models trained on the reference period (f_p):

$$LRI = |f - f_p| \tag{2}$$

Additionally, the LRI is enhanced with a sliding threshold metric based on an average obtained with a rolling-window algorithm. This is done to trigger early warnings while limiting the occurrence of false alarms due to LRI spikes. In particular, as shown in Figure 4, an alarm is triggered for a signal when the following condition is satisfied for P consecutive time steps:

$$LRI_i \geq 0.5 \cdot \frac{1}{W} \Sigma_{j=i-W}^{i} LRI_j \quad (3)$$

where LRI_i is the LRI of the signal at time i and W is the length of the sliding window. The value of W should be selected according to the periodicity of the observed phenomena and, in this specific case, corresponds to 24 h. Thus, as the averaged LRI experiences a deviation $\geq$50% compared to the last 24 h that persists for at least P time steps, an alarm is triggered for the specific sensor of that LRI. We set the persistence threshold P to 6 h, which resulted in effectively removing residual noise.

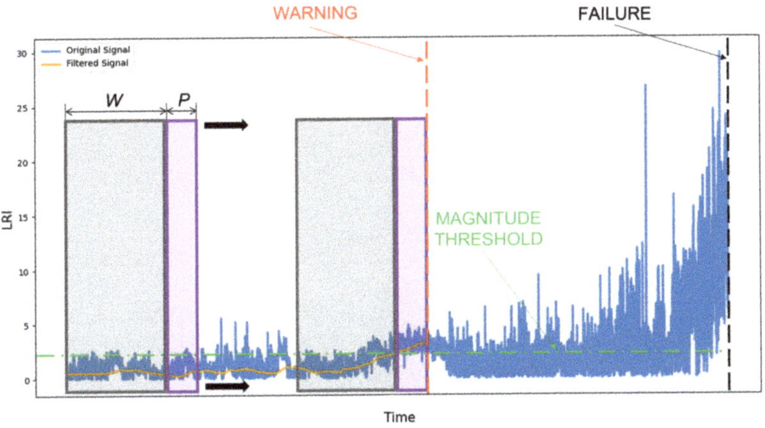

Figure 4. Example of the anomaly detection rule to trigger early alarms on specific sensors.

This approach proved to be particularly suitable for this type of dataset, in which a standard control chart with a fixed threshold for LRIs could be ineffective due to the extreme data variability in some periods and seasons. Moreover, it guarantees high robustness in handling the noise of the residuals of the models.

Based on such a warning rule, model performances were evaluated in terms of anomaly detection capability on each cross-validation datasets cyclically isolated. This assessment aims to quantify the ability of each warning to anticipate the major failure events included in the SCADA log.

4. Results

4.1. Dataset Description

Data were collected from a natural gas genset installed in the Aosta District Heating plant, which is equipped with a 16-cylinders turbocharged ICE. The engine has a nominal electric power output of 7.5 MWe, and it is directly coupled to a 17.5 MWt heat pump. ICE technical specifications are given in Table 1.

A SCADA system monitors different operating parameters collected by the main components of the genset together with environmental measurements. In detail, the initial dataset included 45 parameters sampled every 5 min from September 2019 to December 2020, for a total of 15 months. After the application of the signal preprocessing described in the Section 3.1, the feature number was reduced to 33 significant parameters as listed in Table 2.

Table 1. Technical specifications of the engine.

Quantity	Value	Unit
N. of cylinders	16	[–]
Engine speed	720	[r/min]
Electrical power output	7235	[kW]
Thermal power air cooler HT	1305	[kW]
Thermal power air cooler LT	490	[kW]
Thermal power lube oil cooler	730	[kW]
Thermal power jacket water cooler	925	[kW]
Exh. mass flow rate	39600	[kg/h]
Exh. gas temp.	355	[°C]

Table 2. List of SCADA signals.

Signal ID	Description
P01–P19, P23, P25–P26	Cylinder, exhaust and intake temperatures
P20–P22, P24	Cylinder and fuel subsystem pressures
P27–P31	Generator phase and bearing temperatures
P32	Active power
P33	Ambient temperature

In addition to the SCADA signals, the framework's anomaly detection capability was evaluated by looking at the alerts logged by the SCADA system from October to December 2020, a period when numerous major failures occurred. As described in Section 3.1, only major events were considered, including scheduled (i.e., normal stop) and unscheduled downtime (i.e., emergency stop or outages after engine deratings). Table 3 lists the filtered major SCADA events in the reference period.

Table 3. Event log for major events in the observation period. Event types are abbreviated as follows: D—Derating, NS—Normal Stop and UD—Unscheduled Downtime.

Event ID	SCADA Event Log	Event Type	Start	Duration (hh)
DS_05_10	Exh Temp Deviation Cylinder	D & UD	5 October 2020	11
S_07_10	Emergency Stop Activated	UD	7 October 2020	123
DS_15_10	Exh Temp Deviation Cylinder	D & UD	15 October 2020	11
S_20_10	Emergency Stop Activated	UD	20 October 2020	5
DS_26_10	Exh Temp Deviation Cylinder	D & UD	26 October 2020	11
S_13_11	Emergency Stop Activated	UD	13 November 2020	4
D_16_11	Charge Air Temp After Cooler High	D	16 November 2020	1
S_19_11	Emergency Stop Activated	UD	19 November 2020	48
S_13_12	Shutdown from Main Control	NS	13 December 2020	1
DS_16_12	Generator Stator Temp Windings	D & UD	16 December 2020	1
S_21_12	Emergency stop Activated	UD	21 December 2020	12

4.2. ML Settings and Prediction Errors

Both ML approaches experienced identical training, cross-validation and testing phases. At the training stage, the dataset was split into training and validation sets, respectively, named D_{train} and D_{val}, corresponding to 70% and 30% of the total set. Finally, the testing set D_{test} consisted of a single month cyclically isolated from the available data and included the time periods of failure occurrences.

The XGB model learning task was set to linear regression with hyperparameters optimization according to grid search algorithm, while the MLP setup included early stopping to avoid overfitting. Tables 4 and 5 lists the two subsets of hyperparameters.

Table 4. XGB regressor hyperparameters.

Hyperparameter	Value
Subsampling of columns	0.20
Learning rate	0.10
Max depth	50
Nr. of trees	150
Nr. of parallel trees	20
Alpha	0
Lambda	1

Table 5. MLP regressor hyperparameters.

Hyperparameter	Value
Nr. of Neurons	22
Nr. of hidden layer	1
Nr. of training epochs	150
Activation function	relu
Initial learning rate	1×10^{-5}
Optimizer	ADAM
Batch size	1/50th

The ML model predictions are evaluated in terms of the reconstruction errors of all SCADA signals (during the training phase the predicted values are compared with the actual ones). As can be seen from Table 6, XGB outperforms MLP in terms of customary scores.

Table 6. Reconstruction errors for the proposed ML models.

	XGB	MLP
MAE	0.04	0.11
MSE	0.10	0.14
RMSE	0.21	0.31
MDAPE	0.01	0.13

4.3. *Anomaly Detection Results*

For the evaluation of the anomaly detection capabilities, the results of the testing phase refers to the period of October to December 2020. Specifically, ML model results are discussed by plotting the LRIs against the relative warnings activated on the individual parameters after the application of the sliding threshold metrics (Equation (3)). Furthermore, as a reference to identify engine derating and shutdown, the results are presented in terms of the active power together with the details of the main alarms recorded by the SCADA system in the same time interval.

Figures 5–7 illustrate the results in October 2020. Figure 5 shows the active power, with the detail of SCADA event logs recorded in that period (event IDs refer to Table 3). Figures 6 and 7 show the LRI together with the warnings triggered by the framework (highlighted in dashed red lines).

By analyzing October 2020 SCADA logs, five significant events were isolated. Those events include three anomalies that resulted in a preliminary power output derating followed by engine shutdown, along with two emergency stops linked to unscheduled downtimes. Regarding the first event category, it is worth noting that all the shutdowns were anticipated by cylinder temperature anomalies and that the application of the proposed framework allows for the early detection of such precursors. In particular, for the events detected on 5 October 2020 (event ID: DS_05_10) and 15 October 2020 (event ID: DS_15_10), respectively, a significant deviation of the LRI associated with cylinder temperature parameter (P16) can be seen in Figures 6 and 7, resulting in early warnings with respect to the actual SCADA log (additional details on the advance times relative to the two ML models are in Table 7).

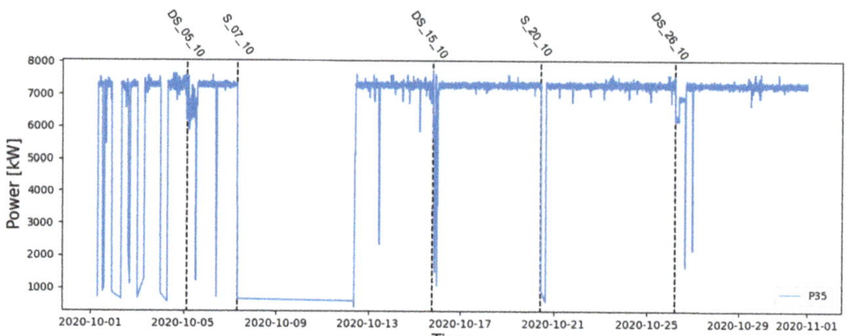

Figure 5. Active power in the reference period of October 2020 with details on the SCADA events recorded in that period (black dashed line).

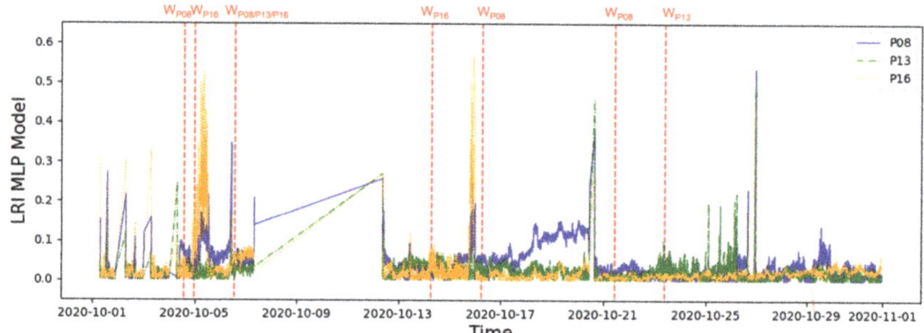

Figure 6. LRIs related to the parameters that caused a warning (red dashed line) after the application of the sliding threshold metric for the MLP model in October 2020.

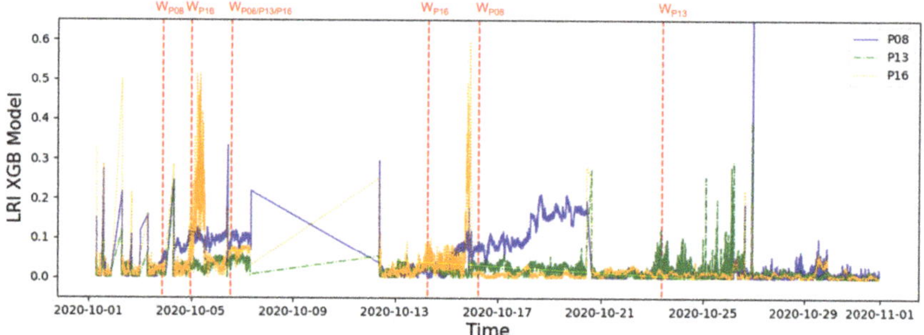

Figure 7. LRIs related to the parameters that caused a warning (red dashed line) after the application of the sliding threshold metric for the XGB model in October 2020.

Furthermore, while the warning on the S_15_10 event was triggered by the two models at the same time, the MLP model detected the anomaly related to event DS_05_10 about ten hours earlier than XGB. The third derating event followed by an engine shutdown was recorded on 26 October 2020 (event ID: DS_26_10) and concerned a high-temperature alarm on cylinder 5B (P13) detected on the same day. Furthermore, for this event, Figures 6 and 7 present a significant variation of the LRI for the parameter P13, constituting a specific precursor that results in a warning both for the MLP and XGB models on 23 October 2020, about three days in advance compared to the SCADA alarm.

Table 7. Comparison of detection performance of unscheduled downtime events (October–December 2020).

Event ID	XGB Results Detection (dd/mm/yy; hh/mm)	Anticipation (hh)	Precursors ID	MLP Results Detection (dd/mm/yy; hh/mm)	Anticipation (hh)	Precursors ID
DS_05_10	4 October 2020; 23:30	4	P16	4 October 2020; 13:45	14	P16
S_07_10	3 October 2020; 20:25	84	P08	4 October 2020; 17:20	62	P08
	6 October 2020; 13:15	18	P13, P16	6 October 2020; 14:05	17	P13, P16
DS_15_10	14 October 2020; 06:40	37	P16	14 October 2020; 08:10	34	P16
S_20_10	16 October 2020; 06:00	101	P08	16 October 2020; 07:05	100	P08
DS_26_10	23 October 2020; 09:30	67	P13	23 October 2020; 10:45	65	P13
S_13_11	10 November 2020; 12:55	69	P08	10 November 2020; 14:15	68	P08
S_19_11	14 November 2020; 00:25	123	P04, P08	14 November 2020; 01:05	122	P04, P08
DS_16_12	4 December 2020; 00:10	299	P28, P29, P30	4 December 2020; 01:25	298	P29, P31
	4 December 2020; 00:20	299	P04, P31, P32	-	-	-
S_21_12	16 December 2020; 12:05	114	P04, P08	16 December 2020; 13:00	113	P31, P32

The advances found before the emergency stops on 7 October 2020 (event ID: S_07_10) and 20 October 2020 (event ID: S_20_10) are of particular interest since they are not associated with a specific SCADA anomaly alarm on a component of the gas genset. In correspondence to these unscheduled downtimes, both ML models showed an anomaly on the LRI of cylinder temperature (P08), which caused a warning three days in advance of the first event (84 h for XGBoost and 62 h for MLP). Subsequently, the indicator of parameter P08 returned to normal values after the maintenance intervention, as visible in the active power plot in Figure 5), and then deviated again from 16 October 2020 (see Figures 6 and 7) until the emergency stop on 20 October 2020.

In a similar fashion, Figures 8–10 compare the results of the CBM method during November and December 2020, during which four significant unscheduled downtimes were reported by the SCADA system. Details on the event log can be found in Table 3.

Those events include three emergency stop alarms recorded, respectively, on 13 November 2020 (event ID: S_13_11), 19 November 2020 (event ID: S_19_11) and 21 December 2020 (event ID: S_21_12) as well as a shutdown transient due to an anomaly found on the generator temperature on 16 December 2020 (event ID: DS_16_12). From a global analysis of the LRI trends, shown in Figures 9 and 10, different anomalies were detected during the observed period in the engine cylinders and generator. In particular, previously found anomalies on the cylinder exhaust temperature, correlated with two long outages in October 2020, recurred from 13 November 2020, when a warning on the involved parameter was triggered by both MLP (Figure 9) and XGBoost (Figure 10). This significant deviation of the P08 parameter indicator persisted for about three days until an emergency stop was recorded on 13 November 2020.

Immediately after this 4-h engine outage, both models detected a new significant anomaly on P08, also involving other cylinders' temperatures and anticipating the failure detected by SCADA on 19 November 2020 (event ID: S_19_11). Of particular interest are the results related to the remaining two significant events recorded by the SCADA in December

2020, namely, DS_16_12 and S_21_12. In fact, the warnings detected so far by XGBoost and MLP were always triggered by the same parameters (with some differences only in the advance times with respect to the SCADA events), while in these two cases, different precursors emerged from the models.

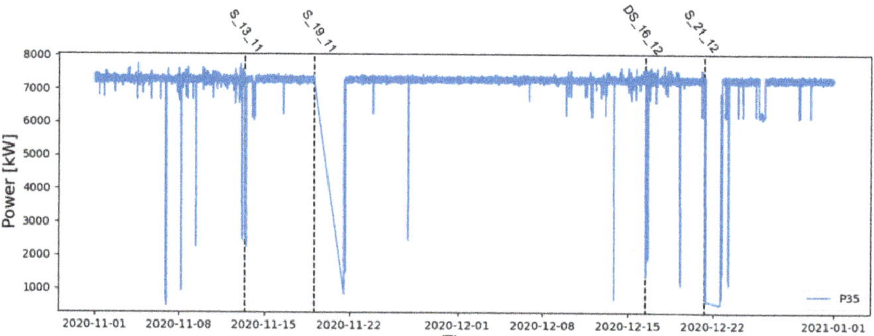

Figure 8. Active power for the period of November and December 2020, with details on the SCADA events recorded in that period (black dashed line).

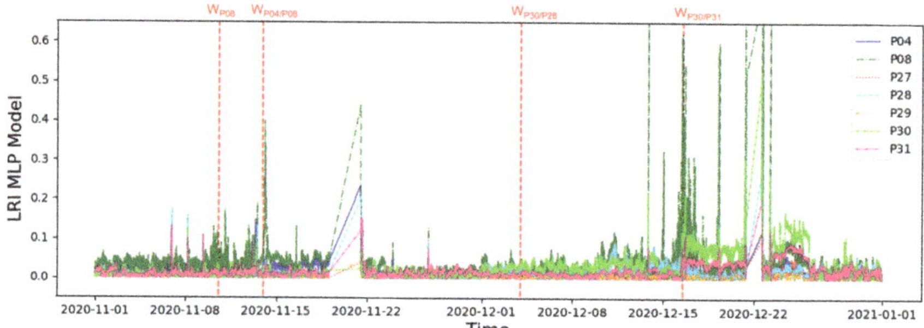

Figure 9. LRIs related to the parameters that generated a warning (red dashed line) after the application of the sliding threshold metric for the MLP model in November and December 2020.

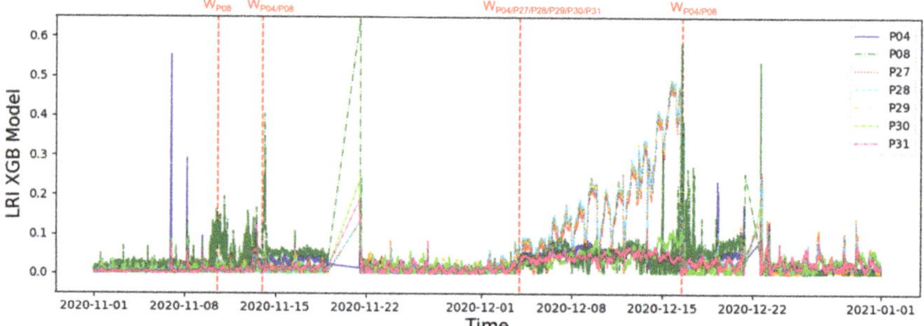

Figure 10. LRIs related to the parameters that generated a warning (red dashed line) after the application of the sliding threshold metric for the XGB model in November and December 2020.

In particular, XGBoost LRIs (Figure 9) highlighted, on 4 December 2020, a variation in the three temperatures of the generator-related variables, phases and bearings (P27-P31). This resulted in a warning that anticipates the SCADA log DS_16_12 by about twelve days. Comparing these results with those of the MLP model (Figure 10), the same significant deviation was not noticed on the generator stator winding but only on the two generator bearings.

As for the unscheduled downtime of 21 December 2020, it was detected about 5 days in advance by both models, with different precursors: exhaust cylinder temperatures (P01-P19) for the XGB model and generator bearing temperatures (P27-P31) for the MLP model.

Finally, Table 7 summarizes the results discussed so far. In particular, the ability of each of the two ML models was assessed to identify specific precursors for major faults included in the SCADA log and then quantified the time of advance warning of the model relative to the occurrence of the reference SCADA alarm.

5. Conclusions

In this paper, an anomaly detection framework for the CBM of natural gas engines used in DH applications was presented. The framework exploited the use of signals collected by the SCADA system. The peculiarities of the framework reside in the PPS-inspired feature selection to reduce dataset dimensionality, the indifference to training dataset clustering to discriminate faults and normal operations and the management of time-series high-frequency information content directly filtering local residuals.

Two different models were tested to represent two different algorithm families: XGB in the symbolist family of decision trees and MLP in the connectivist family of neural networks. These models were trained to learn the regular behavior of the system based on a Leave-One-Out Cross-Validation approach and, based on the model reconstruction errors, a Local Residual Indicator (LRI) was defined for each monitored variable. Therefore, with the aim of triggering an early warning before the occurrence of faults, while limiting false alarms associated with instantaneous peaks in LRIs, a sliding threshold metric based on a moving average was adopted. In this way, a warning was triggered for the signals with the highest reconstruction error, to isolate the parameters mostly involved in the anomaly for troubleshooting purposes.

The proposed method was validated on 5 min SCADA data collected from a 7.5 MWe natural gas engine installed in the District Heating plant of Aosta city. The model was tested on anomalous periods selected using the SCADA event log. The results show that the proposed multivariate nowcasting approach allows the unveiling of hidden precursor dynamics that anticipate all the main fault events that occurred in the observed period. It is interesting to note that these anomalies were not detected by single-variable operational control approaches typical of SCADA systems.

In addition, even if both ML models anticipated the same faults with similar advance times, the better performance of XGB compared to MLP was evident in terms of the training customary scores for the nowcasting of single parameters (see Table 6). In particular, XGB paired with the two-stage threshold tuned with a persistence time of 6 h and time window size of 24 h provided fault anticipations ranging from 4 to 299 h. The framework proved to be fault agnostic because it detected ICE and generator anomalies.

In conclusion, the proposed solution presents a number of benefits due to its nature, which includes the ability to early detect anomalies in NG genset in DH networks, enabling the timely planning of corrective measures before major failures occur. This feature aligns with a CBM approach, where predictive maintenance strategies are adopted to ensure equipment performance and prevent unexpected downtime. Moreover, the proposed solution is cost-effective, as it works directly on the data sampled from the integrated SCADA systems. Unlike other systems that require additional intervention costs, the proposed solution operates directly on the available data and can be seamlessly integrated into the existing system.

The proposed solution employs a non-supervised approach that does not require labels to classify operational states during the training phase, which can be challenging to obtain. This feature makes the proposed solution highly versatile and adaptable to a wide range of systems and contexts. The methodological framework also introduces innovative solutions compared to the state of the art, including a feature selection phase based on CPSS that optimizes the response times of the algorithm to obtain near real-time responses. Additionally, the training approach does not require a preliminary isolation of faulty conditions for the identification of the reference normal behavior model.

Finally, a post-processing of residuals is introduced through the use of a two-stage sliding threshold metric that provides nowcasting of alarms. Overall, the proposed solution offers a highly effective, efficient and cost-saving approach compared to the other systems and methods currently used in the industry. Future research could explore the potential of scaling up the solution for larger DH networks and testing its application in other domains.

Author Contributions: Conceptualization, A.C. and R.S.; methodology, V.F.B. and F.B.; formal analysis, V.F.B.; investigation, V.F.B., F.B. and F.A.T.; resources, R.S.; writing—original draft F.A.T. and F.B.; writing—review and editing, V.F.B. and F.B.; visualization, F.A.T.; supervision, A.C.; project administration, A.C. and R.S. All authors have read and agreed to the published version of the manuscript.

Funding: This reasearch was funded by a contract between Engie Servizi S.p.A. and DIMA, contract number 9149/20. The APC was funded by DIMA.

Conflicts of Interest: The authors declare no conflict of interest.

References

1. Verda, V.; Colella, F. Primary energy savings through thermal storage in district heating networks. *Energy* **2011**, *36*, 4278–4286. [CrossRef]
2. Dominković, D.F.; Stunjek, G.; Blanco, I.; Madsen, H.; Krajačić, G. Technical, economic and environmental optimization of district heating expansion in an urban agglomeration. *Energy* **2020**, *197*, 117243. [CrossRef]
3. Cai, H.; Ziras, C.; You, S.; Li, R.; Honoré, K.; Bindner, H.W. Demand side management in urban district heating networks. *Appl. Energy* **2018**, *230*, 506–518. [CrossRef]
4. Alexandrov, G.; Ginzburg, A. Anthropogenic impact of Moscow district heating system on urban environment. *Energy Procedia* **2018**, *149*, 161–169. [CrossRef]
5. Milić, V.; Amiri, S.; Moshfegh, B. A systematic approach to predict the economic and environmental effects of the cost-optimal energy renovation of a historic building district on the district heating system. *Energies* **2020**, *13*, 276. [CrossRef]
6. Newell, R.; Raimi, D.; Villanueva, S.; Prest, B. Global energy outlook 2021: Pathways from Paris. *Resour. Future Rep.* **2021**, *8*, 11–21.
7. Bouckaert, S.; Pales, A.F.; McGlade, C.; Remme, U.; Wanner, B.; Varro, L.; D'Ambrosio, D.; Spencer, T. *Net Zero by 2050: A Roadmap for the Global Energy Sector*; Report; IEA: Paris, France, 2021. Available online: https://www.iea.org/reports/net-zero-by-2050 (accessed on 13 March 2023).
8. Sayegh, M.A.; Jadwiszczak, P.; Axcell, B.; Niemierka, E.; Bryś, K.; Jouhara, H. Heat pump placement, connection and operational modes in European district heating. *Energy Build.* **2018**, *166*, 122–144. [CrossRef]
9. Levihn, F. CHP and heat pumps to balance renewable power production: Lessons from the district heating network in Stockholm. *Energy* **2017**, *137*, 670–678. [CrossRef]
10. Wang, H.; Yin, W.; Abdollahi, E.; Lahdelma, R.; Jiao, W. Modelling and optimization of CHP based district heating system with renewable energy production and energy storage. *Appl. Energy* **2015**, *159*, 401–421. [CrossRef]
11. IEA. *District Heating*; Report; IEA: Paris, France, 2021.
12. Mendonça, P.; Bonaldi, E.; de Oliveira, L.; Lambert-Torres, G.; da Silva, J.B.; da Silva, L.B.; Salomon, C.; Santana, W.; Shinohara, A. Detection and modelling of incipient failures in internal combustion engine driven generators using electrical signature analysis. *Electr. Power Syst. Res.* **2017**, *149*, 30–45. [CrossRef]
13. Yun, Q.; Zhang, C.; Ma, T. Fault diagnosis of diesel generator set based on deep believe network. In Proceedings of the second International Conference on Artificial Intelligence and Pattern Recognition, Beijing, China, 16–18 August 2019; pp. 186–190.
14. Assuncao, F.d.O.; Borges-da Silva, L.E.; Villa-Nova, H.F.; Bonaldi, E.L.; Oliveira, L.E.L.; Lambert-Torres, G.; Teixeira, C.E.; Sant'Ana, W.C.; Lacerda, J.; da Silva Junior, J.L.M.; et al. Reduced Scale Laboratory for Training and Research in Condition-Based Maintenance Strategies for Combustion Engine Power Plants and a Novel Method for Monitoring of Inlet and Exhaust Valves. *Energies* **2021**, *14*, 6298. [CrossRef]
15. Basurko, O.C.; Uriondo, Z. Condition-Based Maintenance for medium speed diesel engines used in vessels in operation. *Appl. Therm. Eng.* **2015**, *80*, 404–412. [CrossRef]
16. Vera-García, F.; Pagán Rubio, J.A.; Hernández Grau, J.; Albaladejo Hernández, D. Improvements of a failure database for marine diesel engines using the RCM and simulations. *Energies* **2019**, *13*, 104. [CrossRef]
17. Aliramezani, M.; Koch, C.R.; Shahbakhti, M. Modeling, diagnostics, optimization, and control of internal combustion engines via modern machine learning techniques: A review and future directions. *Prog. Energy Combust. Sci.* **2022**, *88*, 100967. [CrossRef]
18. Ntakolia, C.; Anagnostis, A.; Moustakidis, S.; Karcanias, N. Machine learning applied on the district heating and cooling sector: A review. *Energy Syst.* **2021**, *13*, 1–30. [CrossRef]

19. Mbiydzenyuy, G.; Nowaczyk, S.; Knutsson, H.; Vanhoudt, D.; Brage, J.; Calikus, E. Opportunities for machine learning in district heating. *Appl. Sci.* **2021**, *11*, 6112. [CrossRef]
20. Baranowski, J.; Bania, P.; Prasad, I.; Cong, T. Bayesian fault detection and isolation using Field Kalman Filter. *EURASIP J. Adv. Signal Process.* **2017**, *2017*, 1–11. [CrossRef]
21. Flett, J.; Bone, G.M. Fault detection and diagnosis of diesel engine valve trains. *Mech. Syst. Signal Process.* **2016**, *72*, 316–327. [CrossRef]
22. Jung, D. Data-driven open-set fault classification of residual data using Bayesian filtering. *IEEE Trans. Control. Syst. Technol.* **2020**, *28*, 2045–2052. [CrossRef]
23. Czech, P.; Mikulski, J. Application of Bayes classifier and entropy of vibration signals to diagnose damage of head gasket in internal combustion engine of a car. In Proceedings of the International Conference on Transport Systems Telematics, Katowice/Krakow/Ustron, Poland, 20–25 October 2014; Springer: Berlin/Heidelberg, Germany, 2014; pp. 225–232.
24. Zhang, F.; Jiang, M.; Zhang, L.; Ji, S.; Sui, Q.; Su, C.; Lv, S. Internal combustion engine fault identification based on FBG vibration sensor and support vector machines algorithm. *Math. Probl. Eng.* **2019**, *2019*, 8469868. [CrossRef]
25. Dandare, S.; Dudul, S. Support vector machine based multiple fault detection in an automobile engine using sound signal. *J. Electron. Electr. Eng.* **2012**, *3*, 59–63.
26. Tautz-Weinert, J.; Watson, S.J. Using SCADA data for wind turbine condition monitoring—A review. *IET Renew. Power Gener.* **2017**, *11*, 382–394. [CrossRef]
27. Helbing, G.; Ritter, M. Deep Learning for fault detection in wind turbines. *Renew. Sustain. Energy Rev.* **2018**, *98*, 189–198. [CrossRef]
28. Liu, Y.; Chang, W.; Zhang, S.; Zhou, S. Fault diagnosis and prediction method for valve clearance of diesel engine based on linear regression. In Proceedings of the 2020 Annual Reliability and Maintainability Symposium (RAMS), Palm Springs, CA, USA, 27–30 January 2020; IEEE: Piscataway, NJ, USA, 2020; pp. 1–6.
29. Bryg, D.J.; Mink, G.; Jaw, L.C. Combining lead functions and logistic regression for predicting failures on an aircraft engine. In Proceedings of the Turbo Expo: Power for Land, Sea, and Air, Berlin, Germany, 9–13 June 2008; Volume 43123, pp. 19–26.
30. Singh, D.; Kumar, M.; Arya, K.; Kumar, S. Aircraft Engine Reliability Analysis using Machine Learning Algorithms. In Proceedings of the 2020 IEEE 15th International Conference on Industrial and Information Systems (ICIIS), Rupnagar, India, 26–28 November 2020; IEEE: Piscataway, NJ, USA, 2020; pp. 443–448.
31. Maraini, D.; Simpson, M.; Brown, R.; Poporad, M. Development of a Data-driven Model for Marine Gas Turbine (MGT) Engine Health Monitoring. In Proceedings of the Annual Conference of the Prognostics and Health Management Society, Paris, France, 2–5 May 2018.
32. Chen, M.; Li, Z.; Lei, X.; Liang, S.; Zhao, S.; Su, Y. Unsupervised Fault Detection Driven by Multivariate Time Series for Aeroengines. *J. Aerosp. Eng.* **2023**, *36*, 04022129. [CrossRef]
33. Deon, B.; Cotta, K.; Silva, R.; Batista, C.; Justino, G.; Freitas, G.; Cordeiro, A.; Barbosa, A.; Loução Jr, F.; Simioni, T.; et al. Digital twin and machine learning for decision support in thermal power plant with combustion engines. *Knowl.-Based Syst.* **2022**, *253*, 109578. [CrossRef]
34. Braei, M.; Wagner, S. Anomaly detection in univariate time-series: A survey on the state-of-the-art. *arXiv* **2020**, arXiv:2004.00433.
35. Pedregosa, F.; Varoquaux, G.; Gramfort, A.; Michel, V.; Thirion, B.; Grisel, O.; Blondel, M.; Prettenhofer, P.; Weiss, R.; Dubourg, V.; et al. Scikit-learn: Machine Learning in Python. *J. Mach. Learn. Res.* **2011**, *12*, 2825–2830.
36. Kolmogorov, A. On the Shannon theory of information transmission in the case of continuous signals. *IRE Trans. Inf. Theory* **1956**, *2*, 102–108. [CrossRef]
37. Wetschoreck, F.; Krabel, T.; Krishnamurthy, S. Online Repository. 2021. Available online: https://github.com/8080labs/ppscore/releases (accessed on 13 March 2023).
38. Bergstra, J.; Bengio, Y. Random Search for Hyper-Parameter Optimization. *J. Mach. Learn. Res.* **2012**, *13*, 281–305.
39. James, G.; Witten, D.; Hastie, T.; Tibshirani, R. *An Introduction to Statistical Learning*; Springer: Berlin/Heidelberg, Germany, 2013; Volume 112.
40. Goodfellow, I.; Bengio, Y.; Courville, A. *Deep Learning*; MIT Press: Cambridge, MA, USA, 2016.
41. Miele, E.S.; Bonacina, F.; Corsini, A. Deep anomaly detection in horizontal axis wind turbines using Graph Convolutional Autoencoders for Multivariate Time series. *Energy AI* **2022**, *8*, 100145. [CrossRef]

MDPI
St. Alban-Anlage 66
4052 Basel
Switzerland
www.mdpi.com

Energies Editorial Office
E-mail: energies@mdpi.com
www.mdpi.com/journal/energies

www.ingramcontent.com/pod-product-compliance
Lightning Source LLC
LaVergne TN
LVHW070908170726
843515LV00005B/734